Eneas Porto
Marcelo Latuff

Climate seasonality and respiratory diseases

Eneas Porto
Marcelo Latuff

Climate seasonality and respiratory diseases

Barreiras/BA

ScienciaScripts

Imprint

Cover image: www.ingimage.com

This book is a translation from the original published under ISBN 978-613-9-66167-1.

Publisher:
Sciencia Scripts
is a trademark of
Dodo Books Indian Ocean Ltd. and OmniScriptum S.R.L publishing group

120 High Road, East Finchley, London, N2 9ED, United Kingdom
Str. Armeneasca 28/1, office 1, Chisinau MD-2012, Republic of Moldova, Europe
Printed at: see last page
ISBN: 978-620-8-01416-2

"Maybe I didn't manage to do the best, but I fought for the best. I'm not who I would like to be, I'm not who I could be, I'm not what I should be, I'm not what I will be, but thank God I'm not who I was."

(Martin Luther King)

I dedicate this work

To my God.
To my parents, Jercino Rodrigues Porto and Ivone Ana Oliveira Porto, who deserve all the credit for my victory, to my brothers Maxwell and Melquisedeque and to my grandmother Ermesina.

ACKNOWLEDGEMENTS

Words alone would not be enough to express my gratitude to those who were essential to the realisation of this stage in my life, but using them is the least I can do to say thank you.

To God for the strength to overcome every challenge, and to make King David's words my own:

> *Your word is a lamp to my feet and a light to my path (Psalm 119:105).*

To my parents, Jercino and Ivone, for their unconditional love and care, which enabled me to follow every step and taught me that success and humility go hand in hand.

To my grandmothers Joana and Ermesina, for being examples of life and Christian principles, for their words of advice and for their daily and tireless prayers. To the entire Porto Family and the Oliveira Family for their brotherly love and unity.

To my brothers Maxwell and Melquisedeque for their affection and for sharing and contributing the knowledge of their brilliant minds. We know we're going to win.

To the friends I've made during this period, and those of long standing, who have allowed me to experience the true meaning of life: living.

To my neighbours and friends Valdemir and Raquel, and to the Gemelli family for their affection and solid, true friendship.

To our friends from the Academic Directory, Maria Alice, Carla, Lann, Jamires, Emanuela and Solange, for learning together that every day we have to manoeuvre life's obstacles.

To colleagues and former classmates Dayana, Georgitom, Hilton, Heloise, Ianne, Joildes, Lucas, Luís Paulo, Maurício, Marta, Nayara, Patrícia, Suédio, Ueliton, Queurem *(jn memórian).*

Exceptionally, to my supervisor Marcelo Latuf, for the effort and knowledge he passed on, for the shared experience, for every piece of advice and, above all, for his patience in putting up with me all this time.

To professors Pablo Santos, Marcos Mondardo, Paulo Baqueiro, Antonio Lobo, Valney Rigonato, Marcelo de Paula, Mário, Anatália, Evanildo and former professors Elaine, Ricardo, Camila and Gustavo. To the course coordinator Prof Janes Lavoratti, who, respectfully, was a mother, for her dedication to solving the demands within her reach to make this moment possible.

The Federal University of Bahia and the Institute of Environmental Sciences and Sustainable Development (ICADS) for making my training possible. And to others who were part of this achievement.

SUMMARY

The influence of climatic, meteorological and anthropic factors on the health conditions of the population allows geographical science to explore and analyse this relationship, whether through geoprocessing techniques or statistical methods. In Brazil, one of the most prominent problems caused by atmospheric factors are respiratory diseases, the vast majority of which are caused mainly by the unhealthy conditions of the atmospheric environment, such as the significant amount of particulate matter in the air and low relative humidity, especially during hydrologically dry periods. This seasonal dry period that occurs in this region of western Bahia is mainly the result of the action of the dry, hot air mass formed in the centre of the continent, which hinders the occurrence of rainfall and causes periods of scarcity that can exceed 6 months, according to meteorological records from the National Meteorological Institute (INMET). These atmospheric processes favour the worsening of the population's health conditions, especially in relation to the respiratory system. According to the Ministry of Health's DATASUS portal, diseases of the respiratory system are the third leading cause of hospital admissions in the municipality of Barreiras (Bahia). The aim of this study is to analyse the issue in greater depth, investigating the main respiratory diseases and the periods of greatest incidence, based on the data available from health establishments and other bodies in the municipality, as well as climatological data and making an association between them.

Keywords: Spatial analysis; Respiratory tract diseases; Air pollution; Climate seasonality; Barreiras-BA.

SUMMARY

CHAPTER 1

INTRODUCTION

In medium-sized cities, especially those with high population growth, there is a need to understand how natural and man-made factors in the composition of the atmosphere will affect the health of the population. This is intrinsically linked to the fact that these cities are undergoing an intense process of transformation, making it important to plan accordingly.

Health Geography, in terms of climate prospecting, has few studies, most of which are applied to large centres. In the case of Barreiras, studies based on this area of the geographical sciences are still unknown, so it is interesting that work like this is carried out to highlight results that are possibly still lacking.

The relevance of this study is also due to the fact that the area in question is a polarised, dynamic medium-sized city undergoing intense growth, increased by the arrival of new residents, as well as the large circulation of people and vehicles coming from other regions, but mainly from neighbouring cities in search of goods and services.

The city of Barreiras is one of the main cities in the west of Bahia. However, as well as having numerous urban problems, the population is constantly faced with problems related to the atmosphere at certain times of the year, such as low levels of relative humidity, large emissions of suspended elements from fires, dust, particulate matter, aerosols, CO2 and others.

In addition to anthropogenic factors, this is also due to the strong influence of atmospheric factors in the geographical region in which the city is located, such as continentality or the absence of large bodies of water, which can interfere with the supply of humidity, the air masses that act in the region, variations in sunshine, altitude, relief and the climatic characteristics that will act on these areas.

These characteristics result in seasonal weather patterns throughout the year, which will have a direct impact on the atmospheric processes in the area, such as winds and rainfall, allowing the "exchange" or "cleaning" of the air, or even droughts, a period in which people are more susceptible to some climatically vulnerable diseases.

By understanding climate dynamics and the distribution of diseases caused by these elements, it becomes possible to apply recommendations that can be used in possible intervention actions, planning and carrying out disease prevention actions, and become a subsidy for public management in planning to promote health and quality of life.

This study was motivated by the importance of understanding atmospheric dynamics and correlating them with factors that affect people's lives, such as respiratory diseases, in order to generate results and prognoses. However, it should be emphasised that this is not an analysis with a focus on the medical field, but a geographical research project interested in studying the implications of the relationship between man and the environment.

The interest in the subject also arose from the prospect of understanding how diseases occur in the population, as well as curiosity about the seasonal behaviour of the atmosphere in the city of Barreiras and from reading studies related to the Geography of Health.

1.1. Problem

How do the occurrences of respiratory diseases relate to the weather conditions produced by the seasonal climate in the city of Barreiras?

1.2. Hypothesis

The seasonal climate combined with unhealthy air conditions, such as low relative humidity, the emission of particulate matter and aerosol gases are the main preponderant factors for the occurrence of respiratory diseases in the city of Barreiras, especially in the drier months from July to October, when the population is exposed to the most critical conditions.

1.3. Objectives

1.3.1. General objective

The general aim of this research is to correlate the occurrence of respiratory diseases with the seasonal climate throughout the year in the city of Barreiras.

1.3.2. Specific objectives

- Address the atmospheric dynamics of the city of Barreiras in a global and local context;
- Investigate the main respiratory diseases and correlate them with the seasonal climate in Barreiras, comparing these factors on a temporal scale;
- Analysing and linking meteorological data and hospital morbidity from respiratory diseases.

1.4. Characterisation of the study area

Emancipated in 1891, the municipality of Barreiras is located in the far west of Bahia, with headquarters at UTM coordinates E 500608.95 m and N 8657204.50 m, as shown in Figure 1, and has an area of 7859.12 km^2 , in which, according to the 2010 IBGE demographic census, it has 137,427 inhabitants, with the ninth highest population in the state and a population density of 20.41 inhabitants/km.

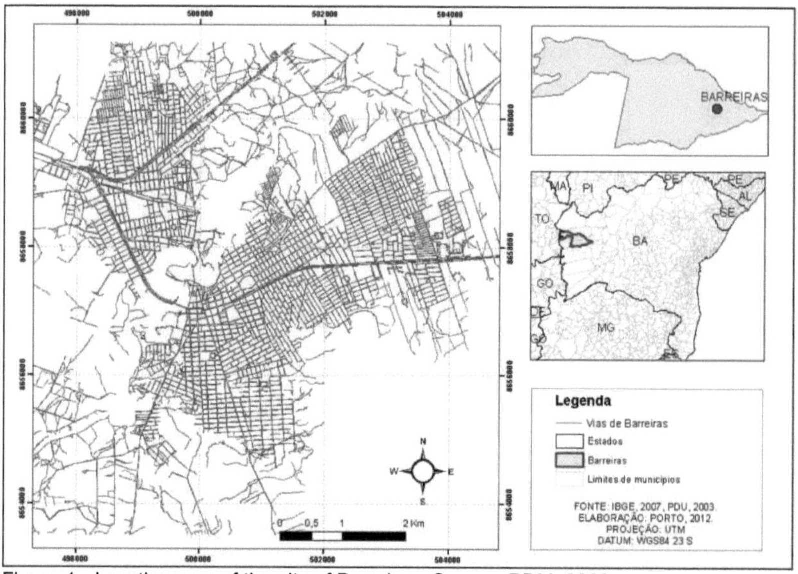

Figure 1 - Location map of the city of Barreiras. Source: PDM, 2003

With an urban population of 123,734 (IBGE, 2010), Barreiras is an urbanised municipality, with more than 90% of the population living in the city, which has formed an urban sprawl of more than 25.97 km^2 in its municipal centre, according to the city's Urban Master Plan (PDDU, 2003).

It is 898 km from the capital Salvador and 645 km from the federal capital Brasília, and stands out as the main urban, political, educational, technological, economic, tourist and cultural centre of the western region of Bahia.

Barreiras is also a regional hub for health services, with 33 public facilities, including the Hospital do Oeste, a regional hospital that serves the whole of western Bahia and even patients from neighbouring states. These demands from neighbouring municipalities lead to a deficit in the capacity and quality of the municipality's health network.

The city is located in a mesoregion with a climate that varies, according to Ayoade (2007), from humid to sub-humid and from dry to sub-humid. Mendonça (2008), on the other hand, goes into greater detail, stating that it is located in the humid-dry or tropical climate of Central Brazil, and in the transition between the climatic subtypes of five to eight dry months.

CHAPTER 2

THEORETICAL FRAMEWORK

Studying the relationship between climate and health from a geographical perspective requires a connection between two aspects. The first involves the link between the characteristics of physical space, such as the environment, the relief, the atmosphere and conditions on a time scale, such as the seasons, seasonality or the behaviour of the climate. The second aspect involves the individual's relationship with the environment, taking into account their way of life, environmental conditions, housing, work and, above all, quality of life.

When working with the conceptualisation of the subject, it can be understood that questions about the quality of life of the population have long been present in the relationship between the environment and individuals. However, it is necessary to review studies and theories that have evolved over time, along with the emergence of new issues.

This chapter will cover some important concepts that range from broad themes of Health Geography, with a historical survey of the emergence of this branch of Geographical Science, to more specific subjects, such as meteorological parameters and some approaches to respiratory system diseases (RDS).

2.1. Geography of Health

The approach to human health as a consequence of geographical factors has been worked on since antiquity with Hippocratic studies, according to Neto (2007), between 460 and 370 BC, a period in which medicine had not yet made great strides and was only informative in nature, with no in-depth analysis.

From there, one of the foundations of scientific medicine began, proposing careful observation of the physical environment in which the illness occurred, the place, the season, the state of the atmosphere and other relationships (SOUZA, 2008). For Hippocrates, the climate exerted a therapeutic influence on human beings (JESUS, 2010), an idea that was greatly developed in one of his great works *Airs, Waters and Places.*

The emergence of studies on the association between geographical elements and diseases, based on Hippocratic studies, provided a new tool for knowledge and action for medical science. However, according to Junqueira (2009), the greatest development in this area occurred after the penetration of imperialist countries into the tropics, where knowledge about the diseases related to these regions became essential for their advancement.

The rapprochement between medical knowledge and Geography only began

> in the 16th century with the great discoveries, which made it necessary to know about diseases in the conquered lands in order to protect the colonisers and develop commercial activities. This period corresponds to the predominance of Geography's deterministic conception of the relationship between man and nature, so that geographical characteristics, especially the climate, were held responsible for the occurrence of diseases (COSTA E TEIXEIRA, 1999, p. 273).

Even with these discoveries, there was still a great deal unknown about many climatic factors and the behaviour of the Earth's atmosphere.

Even so, there were technical developments in the instrumentation of weather observations and in the transmission and analysis of meteorological data, according to Ayoade (2007), which played a vital role in the progress of modern meteorology and climatology, with an intense search for new scientific perspectives. As stated by Costa and Teixeira (1999):

> This movement favoured the birth of tropical medicine, a medical speciality that adopted the idea that some infectious and parasitic diseases were specific to one part of the globe, the tropics, where the hot climate would weaken the human organism, exposing it to these diseases (Pessoa, 1978). During the following two centuries, the medical literature was dominated by eminently descriptive works, emphasising the influence of the environment on man.

These studies were called Medical Geography and presented maps containing information on precautions that should be taken by European armies in the event of military occupation of the "unhealthy" territories of the tropical world. For Ayoade (2007), the rapid development of atmospheric science had to wait for the technological revolution of the Renaissance period. This allowed for the creation of new conceptions of the natural cause of diseases and their association with the climate.

In the 18th and 19th centuries, information on the spatialisation of diseases was systematised (PESSOA, 1978, cited by JUNQUEIRA, 2009). With the results of these systematisations, medical scholars, maintaining a deterministic view of cause and effect, attributed diseases to the physical environment. In a sense, the hygienism developed in the 19th century revived the Hippocratic ideology, which brought back to modern times the importance of considering the conditions of ventilation and adequate lighting in homes in urban planning (MARTINS, 1994).

The Bacterial Theory, created at the end of the 19th century with the discovery of bacteria and parasites, led to a reformulation of the concept of the health/disease process.

Thus, medicine began to focus on the interior of the human body, where it lost interest in the environment, which led to the exclusion of medical geography from medical curricula (LIMA NETO, 2000, quoted by PEITER, 2005). This led to a long decline in medical geography studies.

During the Second World War, Medical Geography once again became important on the medical scene, as knowledge of diseases on the battlefields became fundamental and strategic, according to Junqueira (2009).

Studies that emphasised the impact of the environment, especially the climate, on human health conditions were once again produced and valued (PYLE, 1792 cited by COSTA E TEIXEIRA, 1999). But the most accentuated growth only happened from the 1950s onwards, when there was greater joint organisation of the area. In this sense:

> According to Peiter (2006), it wasn't until around 1949 that Medical Geography was revalorised, which, according to Lima Neto (2000), should be attributed to the fact that it was placed as a branch of Geography, since until that time it was produced predominantly by doctors (ROJAS, 1998). Also according to Freitas (2001), Medical Geography was only officially recognised at the Lisbon International Congress, thanks to the influence of the definition of health presented by the World Health Organisation (WHO) in 1947 (JUNQUEIRA, 2009, p. 4).

According to the same author, political and social changes began to require Geography to make a deeper contribution to the relationship between health and disease. In order to respond to this need, a new branch of this science, Health Geography, emerged in 1976 in Moscow. Its name was changed at the request of the UGI (International Geographical Union) Medical Geography Commission, due to the broadening of the themes, issues and approaches it had developed over time.

The request and justification for changing the name of this branch of Geographical Science came about because Health Geography is considered to be more comprehensive, relating to quality of life, education, housing, basic sanitation, health infrastructure and others. However, this new name is not used in all countries, and the term Medical Geography is still used today (ROJAS 1998; LIMA NETO, 2000; PEITER, 2005).

In Brazil, the first studies related to the distribution of diseases were carried out before the 18th century, in the interests of the colonisers and the court, according to Morais (2007). However, it was from this century onwards that foreigners and naturalists carried out studies that emphasised diseases related to the physical aspects of the environment.

In the following century, according to Junqueira (2009), with the foundation of the country's Faculties of Medicine in 1808, various works appeared that addressed issues relating to the Geography of Diseases or Geographical Pathology of the various Brazilian regions with a deterministic slant.

After a great decline in scientific production in the country at the end of the 19th century and the first half of the 20th century, it was only in the 1930s that pioneering studies appeared in the country, such as the work of Afrânio Peixoto (1938), who established a relationship between some diseases and the country's climatic conditions, in which he demythologised and established the definitions of climatic or tropical diseases.

Ferreira (2003), quoted by Junqueira (2009), states that from the 1950s onwards, geopolitical interest in the interiorisation and national integration of Brazilian territory encouraged studies.

This process resulted in a lot of research into so-called tropical diseases, which were present in these regions, especially in the Amazon and the Centre-West. The same author also reports on the publication of Lacaz (1972), the book Introduction to Medical Geography in Brazil, which has contributed to the historical recovery of discussions between Geography and health.

It was in the 1970s, with the creation of the National School of Medical Geography, that medical geography was consolidated, which, according to Rojas (2003), after the contribution of critical geography, came to be called health geography.

With a new epistemological approach, Milton Santos contributed to redefining and changing conceptions, and to the formation of this new character, which prioritises analysing the interaction between the social environment and the environment based on Critical Geography, as well as defining a geographical concept as the object of study for working on diseases, space.

It was through the Marxist approach and the social analysis of space that Santos effectively contributed to the work of Health Geography (JUNQUEIRA, 2009). Space is therefore Santos' (1980) proposal for analysing the relationship between public health and geographical aspects,

Geography has had great difficulty in studying the health phenomena of human collectivities, as it is characterised as a synthetic discipline that encompasses the human sciences as well as the natural or physical sciences.

This combination of synthesis and imprecision led to a consequent delay in the definition of its object of study, which also hindered its acceptance in academic circles.

However, the race to find this fundamental element for science took place quite efficiently, as soon as this need was realised. Thus:

> Seeking a scientific character, geographers initially relied on the model of the natural sciences (Vasconcelos, 1995), based solely on contemplating the physical nature of space, which in its broadest and most deterministic sense corresponds to the entire earth's surface and biosphere (COSTA E TEXEIRA, 1999).

The systematisation of this field of knowledge only took place at the beginning of the 19th century in Germany, when the search for the object of study of Geographical Science began. This search took place through an intense process of theoretical reformulations in Geography, generating major changes and crises in the currents of thought of this science. In this sense, Santos (2008) emphasises that:

> The demands arising from this process, combined with the internal weaknesses of the discipline itself, such as the vagueness of its subject matter, contributed to triggering a crisis in geography, as its assumptions increasingly became insufficient to respond to and explain the changes that

were taking place.

He also points out that in this context, while maintaining the dominant conceptual bases, the movement known as Pragmatic Geography emerged, which presented them with a technological aspect, materialised through Quantitative Geography and Systemic Geography.

However, it wasn't until the 1970s that a new dominant movement emerged with radical opposition that went beyond academic boundaries, with an intense socio-economic debate, explicitly incorporating politics into scientific discourse and establishing its roots in social issues.

With the emergence of Health Geography in Brazil, the geographical concepts proposed by Milton Santos are one of the most important references for analyses of the relationship between space and disease, especially those produced in Brazil, says Czeresnia (2000).

Space comes to be seen as the result of the dynamics of its complex organisation and interactions, including all elements, including the physical (Santos, 2008), becoming the object of study for Geography. This is conceived as geographical space humanised by social relations and was effective as a category of analysis in bringing medical knowledge and geography closer together. Thus, human space is necessarily the product of a series of decisions that guide its organisation.

Space as a unit of analysis allows for a better understanding of the occurrences and factors resulting from processes related to climatic seasonality and respiratory diseases.

The organisation of urban space also has a direct impact on health, and as this is produced by social agents through different land uses juxtaposed to each other, and in an unequal way, so will public health. For Correia (1993), when we look at the intra-urban area we find a fragmented, articulated city, a reflection and social conditioner, the city is like the place where the various social classes live and reproduce.

Thus, the creation of peripheral spaces, which do not have easy access to public health services, where the majority of people on low incomes are among the most vulnerable groups.

Steinberger (2001) states that the construction of a dignified life in cities, by expanding the rights of its inhabitants, can be established on the basis of the environmental quality of urban life.

However, the growth and development of urban space does not take place homogeneously, but produces privileged areas and segregated areas, which although fragmented are articulated through the complexities of land use. Thus:

> They define areas such as the city centre, where commercial, service and management activities are concentrated, industrial areas and residential areas, distinct in terms of form and social content, leisure areas and, among others, those set aside for future expansion, (CORRÊA, 1995, p.1).

The production and reproduction of urban space is part of a complex system based on demands, interests and administrative and/or political decisions, and often the managers of space, especially the state, are unable or even uninterested in monitoring, organising and ordering urban demands and the growth of cities.

However, the quality of urban life is provided for in the guidelines of the Urban Development Master Plan (PDDU), through the full development of citizens' right of access to housing, urban services and facilities, such as basic sanitation, electricity, tree-lined streets, health, education, culture, leisure and recreation.

One of the basic instruments of a municipal planning process, the Master Plan, can be defined by Brasil (2002) as a set of principles and rules that guide the actions of the agents who build and use urban space.

2.2. Public Health

When people talk about public health, they almost always allude to the occurrence of diseases or even disease treatment systems, but its definition is much more complex and has a very broad meaning. According to Junqueira (2009), the World Health Organisation (WHO) defines health as the state of complete physical, mental and social well-being and not just the absence of disease or infirmity.

This means that issues involving public health are not restricted to the internal behaviour of the human organism or to clinical and hospital environments, but are directly linked to socio-cultural, economic and political aspects, with the life characteristics of the individual, the geographical space where they live, the quality of the environment, the level of education and all their relationships with the environment.

In order to approach or analyse the state of health of a population, numerous points must be considered, both quantitative and qualitative, given their complexity. Thus:

> In order to be healthy, people need an environment that provides them with a state of complete satisfaction, including food, housing, work, sanitation, recreation and disease prevention (MOTA, 2006 p. 41).

This implies that in order to achieve a better state of health, improvements in quality of life must also be sought, where these two aspects have parallel meanings and one directly interferes with the other.

Making this association, the WHO quality of life assessment tool, *WHOQOL*[I] *(World Health Organisation Quality of Life,* 1995) states that quality of life can be defined as an individual's

[I] *WHOQOL - World Health Organisation Quality of Life*

perception of their position in life, in the context of the culture and value systems in which they live and in relation to their goals, expectations, standards and concerns.

This approach to quality of life only takes into account human relationships, social issues and behaviour, which only refers to material, socio-economic and psychological well-being.

Menezes (1981), on the other hand, goes further in his approach to quality of life, using it as a link between health and socio-spatial aspects. He states that:

> It is a notion associated with the environment, in the sense that it is the result of the factors that pressurise and interact with the community. Depending on how you look at it, quality of life is also called standard of living in economic terms, living conditions in social terms, environmental conditions in ecological terms. Quality of life results from the conditions of food, water supply, housing, transport, sewage and rubbish disposal, employment, income, schooling, security, air quality, leisure, hygiene and safety at work, etc.

It is clear, then, that the relationship between human health and quality of life must go beyond isolated actions, having a direct link between local and global spheres that adjust to the dynamic balance of ecosystems, reflecting both the health of individuals and communities. In this sense:

> Health is the main element in sustainable development, where the construction of urban space should aim not only to improve the quality of the physical space (built and circulation), but also the ability to project its consequences on the quality of life and health of individuals and communities, (SANTANA, 2006).

Given the importance of public health in urban development, it is necessary to reflect on practices and objectives so that it can make a greater contribution to quality of life. From this perspective, Cezarina (2006) works on the concepts of disease prevention and health promotion, raising a multidimensional perception related to economic, political, social, cultural and other issues.

From this perspective, the public health system is fundamental for managing and organising health actions to combat and treat diseases, through strategies that guarantee the population basic conditions of care to welcome individuals exposed to illnesses, with quality public services.

Brasil (2012) highlights the 1988 Federal Constitution, which determined that it was the duty of the state to guarantee health to the entire population and, to this end, created the SUS (Unified Health System). Another important step for public health in the country took place in 1990, when the National Congress approved the Organic Health Law, which details how the system works.

Even though it is well standardised and regulated, the public health system in Brazil faces serious problems in all its spheres, ranging from global aspects, such as inefficiency in the administration of resources, to local ones, such as inefficiency in care, which end up directly affecting the quality of the system, with major impacts on the lives of users and the scrapping of the public network.

The SUS is a strategic management model for a service that is essential to

> a large section of the Brazilian population. However, one of the problems in implementing the model proposed by the SUS is undoubtedly the misinformation and lack of knowledge on the part of those who govern. (RIBEIRO, 2006)

It's clear that in order for the administration to function properly, people need to be trained and aware of all the information about the sector in question. Ribeiro (2006) states that the Brazilian Centre for Health Studies (CEBES), in a dossier prepared in 2010, points to even more serious problems, such as the diversion of health resources for corruption and the use of public services as an electoral tool.

These impacts have a direct impact on access to the public health service, where those who don't have private plans have no alternative but to wait for the slow and precarious public service.

With this reality, many patients only seek care when they are in more critical situations. This means that many patients do not seek medical attention and are not quantified in SUS records, thus being left out of statistics and analyses. Yunes (1999) states that this discrepancy occurs mainly with individuals considered to be less vulnerable, such as those in the young-middle age group.

The quantification and collection of SUS data is carried out by DATASUS, a computerised system run by the Ministry of Health, created to provide transparency and information on public health resources. This system contains a database that provides information that can be used to support objective analyses of the health situation, evidence-based decision-making and the development of health action programmes.

According to Brasil (2012):

> Hospital units participating in the SUS, whether public or private, send information on hospital admissions via the AIH (Authorisation for Hospital Admission), to municipal managers (if they are fully managed) or state managers (for others). This information is consolidated in DATASUS, forming a valuable database containing data on most hospital admissions in Brazil.

In the construction of Health Indicators, outpatient and morbidity data, disability, access to services, quality of care, living conditions and environmental factors have come to be used as metrics, which translate into relevant information for quantifying and evaluating health information.

2.3 A geographic approach to climatology: climate and meteorological parameters

In order to understand climate-related health problems, some fundamentals of climatology will be covered, with an emphasis on atmospheric factors and meteorological elements such as temperature, precipitation, wind, humidity and air pressure.

Conceição (2000, p. 13) emphasises the importance of this theoretical approach by quoting Santos (1990), who states that there has always been, and still is today, much more discussion about Geography than about Space, which is the object of geographical science. In the same way, when trying to produce a Geographical Climatology, one must concentrate on the concepts of climate.

The discussion should also be from an epistemological point of view, since Climatology is a science, and according to Mendonça (2007), it is focused on the study of the spatialisation of atmospheric elements and phenomena and their evolution.

When considering the influence of climate on atmospheric behaviour, it is necessary to explore and understand its dynamics and characterise its influence on atmospheric weather, as well as on the study area, so that only then can the intensity of the relationship between these elements and man be understood.

Understanding them is essential for health-related studies, given the relevance and intensity of the influence they have exerted on environmental conditions, human activities and directly on each individual's body. For Ayoade (2007), climate is perhaps the most important component of the natural environment, influencing man and making him dependent on climatic elements for survival, such as air, rain and others.

According to Conceição (2006), Geographical Climatology is the study of how atmospheric dynamics act in a given fraction of Geographical Space.

This leads us to understand that the geographical analysis of climate must start from a global situation, i.e. the large atmospheric systems that act on the macroclimate scale, in order to understand the dynamics of local behaviour, the microclimate.

Thus, larger phenomena and factors that have a direct impact on climatic behaviour would be just as relevant as the factors in the space being cut out.

Even with the concept of space that has already been worked on here, it is worth revisiting the concept of Milton Santos (1996, p. 51), whose definition says that it is formed by an inseparable, solidary and also contradictory set of systems of objects and systems of actions that are not considered in isolation, but as the single framework in which history takes place.

From these ideas it is clear that climate concepts can be worked on here with a focus on Geography, with the creation of a personality in the branch of this science that studies climate, characterising it as Geographical Climatology.

First of all, it is necessary to define the standards used to characterise the climate of an area or region, because it is from these standards that the dynamic processes that act on them can be understood.

According to Ramos (2009), in 1872, the International Meteorological Committee decided to compile values for a uniform period, recommending a 30-year period. According to the same author, in 1989 the World Meteorological Organisation (WMO) recommended updating the 30-year normals every 10 years.

The WMO has therefore come to define normals as average values calculated over a relatively long and uniform period, comprising at least three consecutive decades.

In the event that recent station normals are not updated, or for some other reason, the WMO recommends using provisional normals. Provisional normals are short-period averages, based on observations extending over a minimum period of 10 years.

All the elements, factors and reactions present in the Earth's atmosphere require a primordial component: air. According to Mendonça (2007), the air we breathe is not a homogeneous substance, but a compound of gases, always moving in the atmosphere, and behaves in an atmospheric dynamic that is influenced by countless climatic factors.

Cavalcanti (2009) states that the Earth's climate is defined by two fluids: air and water, each of which has its own dynamics and interacts with each other. However, this approach does not negate air as the foreground in climatology, but only gives importance to the water component due to its significant influence, especially in relation to ocean currents.

Ayoade (2007) states that atmospheric processes formed from the circulation of this air influence other parts of the environment, mainly the biosphere, hydrosphere and lithosphere, which then modify the behaviour of the atmosphere, generating a continuous process and, according to him, imprinting particular characteristics on the behaviour of each place from these interactions, as well as continuously exchanging energy and matter with each other.

Therefore, when analysing the climatic processes in certain areas, other parts of the environment cannot be ignored, since they can cause changes, directly or indirectly, in this cycle.

The air, from its concentration in large packages, forms the air masses, which will control the behaviour of the weather in the earth's atmosphere, imprinting a normal or standard meteorological behaviour wherever they pass, where the great climatic zones emerge.

An air mass is defined by Hare (1963) as a large, horizontal and homogeneous body of air, moving as a recognisable entity and having both tropical and polar origins, Ayoade (2007, p.99). These masses originate in areas with conditions that favour the uniform development of vast bodies of air, such as large bodies of water, wide areas of flat terrain or even extensive depressions.

According to the same author (2007, p. 101) air masses are very important in the study of weather and climate, as they directly influence the area in which they predominate.

The boundaries that separate air masses, i.e. the border zone between masses with different properties, are called fronts. Fronts are mobile and dynamic, and move according to the movement of air masses, which is almost always accompanied by meteorological events.

As they move from their regions of origin, from which they acquire their main

> thermohygrometric characteristics, air masses influence the regions they pass through, bringing new temperature and humidity conditions to these areas and at the same time being influenced by them, (MENDONÇA 2007, p.100).

These movements are mainly responsible for the rains and droughts in many Brazilian regions, forming different types of weather, especially in areas further from the coast, which experience long periods of drought during the year, where there is less influence from the sea.

For Monteiro (1971) quoted by Conceição (2008), the basic attribute in a region like the Brazilian Northeast is rhythmic irregularity, which is responsible for the extreme oscillations that produce prolonged droughts and rapid and equally catastrophic floods.

The same author (1971, p. 6) mentions that the first valid approximation for the concept of rhythm would be that of the annual variations perceived through the monthly variations of climatic elements, whose repetition would be the foundation of the notion of regime.

In this way, the normal behaviour in these areas, which is vulnerable to anomalous variations, is generated from the constant movement of these masses and fronts in the atmosphere, which result in weather conditions and the determination of the local climate.

In terms of climate, what is called "normal" is nothing more than an average, according to Conceição (2008). Climatological normals serve as a reference, an initial characterisation, whose representativeness is only known through the extreme values and the usual rhythm of the sequences of weather types and their geographical implications.

Thus, the action of the main air masses that influence Brazil directly interferes with atmospheric behaviour and the formation of climatic processes in the study area.

The cEm (Continental Equatorial Mass) is basically formed as a result of low pressure in the Amazon region, i.e. it forms over the heated continent, where lulls and weak winds dominate, especially in summer (TORRES E MACHADO, 2011).

In the western region of Bahia, this air mass exerts a great deal of influence on the rainfall regime, especially in the months of December to February, and causes an increase in rainfall heights and air humidity percentages.

In summer, this mass adds strength to the mTa (Atlantic Tropical Mass), which forms in the centre of subtropical high pressures in the Atlantic, where it receives heat and humidity from these warm maritime regions. It acts all year round on the Brazilian coast, but only gains strength in the interior of the continent in winter due to the retreat of the mTc (Continental Tropical Mass).

At the end of the summer, the mPa (South Atlantic Polar Mass) moves along the country's coastline from the South until it reaches the Northeast. On this path, this mass splits into branches, with one part heading towards the south-east of the country and the other towards the centre-west, where it also joins forces with the mTa and reaches areas of the central plateau that covers part of the states of Tocantins, Bahia and Goiás.

Formed between winter and spring in the depression of the South American continent as a result of the intense warming of the surface, the TCM is a pocket of hot, dry air that has a major influence on the central region of Brazil, as it moves across the continent, interacting with air from other locations such as the Centre-West and the interior of the Northeast, mainly between the months of August and October.

According to Torres and Machado (2011), due to the low humidity content of the air, cloud formation and precipitation do not occur. As a result, the days are very sunny, with intense daytime heating and intense night-time cooling.

This is the most critical period for people living in this region, where the unhealthy weather conditions, such as low relative humidity, cause them to feel fatigued and unwell, which can weaken the body.

Fronts are responsible for redistributing heat, reducing the excess in the tropics and the deficit at the poles. Cavalcanti (2009) classifies them as warm fronts and cold fronts:

> Cold air from the poles and warm air from the tropics are separated by what are known as fronts. When the cold air moves towards the equator, the front is called a cold front; when the warm air moves towards the poles, the front is called a warm front.

The masses act as an "engine" that provides cyclical movement in the atmosphere, varies in intensity and causes periods of drought and rainfall at different times of the year, depending on the area of influence, and varies spatially according to the forces acting on them, such as solar radiation, ocean currents and relief, and also involves other elements and climatic factors.

Climate phenomena and climate types can thus be seen as a result of these relationships.

According to Mendonça (2007, p.41), a better understanding of the various types of weather and climate is based on climatic elements and climatic factors (or geographical climate factors). For him:

> Climatic elements are defined by the physical attributes that represent the properties of the geographical atmosphere of a given location and the most commonly used to characterise the geographical atmosphere are temperature, humidity and pressure. Climatic factors, on the other hand, correspond to the static diversifying geographical characteristics of the landscape, such as latitude, altitude, relief, vegetation, continentality, maritime conditions and human activities.

Influenced by geographical diversity, the elements of climate manifest themselves through atmospheric weather behaviours such as precipitation, relative humidity, wind, cloudiness, cold-heat waves and others.

The time-space variation in the manifestation of these elements depends on the actions of climate controls that occur from the combination of climatic factors with the dynamic aspects of the oceanic and atmospheric environment, such as ocean currents, air masses and fronts, which, acting together, create different climates on Earth.

Within this concept, although each of the elements of the climate has a specific dynamic, they should not be considered separately from the climatic factors. This is because, according to Mendonça (2006), the elements act on each other in a significant way, and one element may be active in controlling the other.

The list of elements involved can be considerably expanded depending on the climate analysis you want to carry out, since the interactions between climate elements and factors are individualised in various fields, such as the thermal field (which involves temperature), the hygrometric field (which involves air humidity), the pluviometric field (involving precipitation) and others.

It is therefore necessary to address the continentality effect as one of the climatic factors that has a major influence on the mesoregion of Western Bahia, since these are especially manifested in temperature and relative humidity.

> The continentality of a place is given by its distance from the oceans and seas, which no longer directly exert the actions presented above. In the absence of the mitigating effects of the oceans on temperatures, the heating/cooling of continental surfaces occurs more quickly and with less participation from the humidity of the air, so that, as well as being drier, these places have marked daily thermal amplitudes, (MONTEIRO, 2006, P. 49).

The effects of continentality on climate seasonality are very evident in the Far West of Bahia, since for much of the year, cold and humid air masses lose strength and are unable to reach this region, as well as a large central portion of the country, and breezes from the ocean are prevented from entering these areas, resulting in dry periods and very low cloud cover.

This dry air directly influences the temperature, as well as the daily temperature range, which is mainly high during the day and often drops sharply at night. With the scarcity of clouds, the daily periods of sunshine are longer and more intense, directly affecting the temperature range.

The thermal field is formed by the various temporal and spatial temperature conditions. For Ayoade (2007), temperature can be defined as the movement of molecules, so the faster the movement, the higher the temperature.

Physics explains that the factor that determines a body's temperature is the balance between incoming and outgoing radiation and its transformation into latent or sensible heat. In other words, the solar energy that hits the earth's surface is transferred to the air and transformed into heat.

According to Mendonça (2007), air temperature is the average sensible heat stored in the air. It is commonly measured by thermometers and given in degrees Celsius or Fahrenheit. It can also be measured in Kelvin or in absolute temperature scales, but in Brazil it is customary to use a scale in degrees Celsius (International System of Measures).

Air temperature can vary both spatially and temporally, with daily, monthly or annual changes, or with air circulation and displacement. The difference in demonstrating each of these is that the variation on a temporal scale is simulated on a graph, while the distribution of temperature in space is usually represented by means of isothermal lines.

Ayoade (2006, p.52) states that various factors influence the distribution of temperature over the Earth's surface or part of it. The main factors responsible for spatial variation include the amount of insolation received, the nature of the surface, bodies of water, prevailing winds and ocean currents.

Altitude and latitude also exert great control over temperature, since temperature decreases as altitude increases, as does latitude, where the decrease in temperature is towards the poles, starting at the equator, and according to Ayoade (2006), this is evidence of the level of insolation in the different global zones modified by the angle of the solar inclination.

The different levels of insolation are mainly due to the location of the continental and hydric surfaces and their position in relation to the sun. Thermal conditions in the atmosphere do not only vary horizontally, but there is also vertical variation in air temperature.

> The temperature normally decreases with increasing altitude at a rate of approximately 1°C/100m (dry air adiabatic gradient). This rate of cooling occurs because a rising mass of dry air is subject to lower and lower pressure, increasing its volume and decreasing its temperature, (FRITZSONS, 2008 p.49).

Thus, higher places tend to be colder than lower altitude points, such as valley bottoms. Mendonça (2007) states that there are situations, however, which cause this behaviour to be reversed, generating the phenomenon of thermal inversion, i.e. instead of cooling with altitude, the air starts to warm up.

In the west of Bahia, this occurs preferentially in winter, when there is no wind or cloud cover, causing the soil to lose energy rapidly through radiation, transmitting heat to the overlying air layers. When the air is colder and consequently heavier, it tends to descend, occupying lower parts, such as valley bottoms.

In this way, they become especially harmful when they occur in urban and industrial areas, because they tend to hinder the dispersion of pollutants generated by the activities that take place there, and also intensify the magnitude of heat islands.

Seasonal variations in temperature increase with latitude, but even at relatively low latitudes, the degree of continentality can make this variation very pronounced, causing temperatures to become higher in summer. For Ayoade (2007), these seasonalities are mainly the result of seasonal variations in the amount of insolation received anywhere on the globe.

This same process, albeit on a much shorter time scale, produces the diurnal variation in temperature, which has a greater amplitude at low latitudes. This is because, in these same areas, the daily variation in the Sun's elevation is great, especially in winter.

In winter, in the interior of the Northeast, the daily temperature range is greater, because the low humidity combined with the absence of clouds makes the day very hot and consequently fatiguing, and at night the dry air loses heat very quickly, resulting in a sharp drop in temperature. This process results in weather conditions similar to those in the desert, with high temperatures during the day and low temperatures at night.

In large and medium-sized cities in the interior of Brazil, especially during the winter, the daily thermal variation combined with the gases and particles emitted into the urban atmosphere cause the formation of a spherical layer that mainly surrounds the central areas, where there is intense verticalisation and low wind circulation.

> In an urban area, in addition to the natural components present in the troposphere, polluting or non-polluting gases and particles are released into the atmosphere which, under different meteorological conditions of pressure, temperature, humidity and solar radiation, undergo reactions forming additional pollutants to those already emitted, which contribute to worsening pollution in urban centres (CASTANHO, 1999).

This layer behaves like a dome and prevents the dispersion of smoke, gases and particulate matter present in urban areas, leaving the air dense and heavy.

Solid or liquid particles emitted by sources of air pollution or even those formed in the atmosphere are called particulate matter and, when dispersed in the air, form so-called aerosols (ALMEIDA, 1999).

At night, the emission rate is generally lower, but this dome tends to contract due to the decrease in convective currents, because of the loss of air temperature, which makes the urban atmosphere denser and with a higher concentration of pollutants.

This situation is aggravated by the low percentage of water in the atmosphere, with these values, in the driest periods, reaching very close to zero.

According to Ayoade (2007), water vapour in the atmosphere represents only 2% of the total mass of the atmosphere and 4%, but it can reach zero depending on its distribution in space and time. Like temperature, various factors influence the conditions of the hygrometric field, but it is worth emphasising that this is one of the most important atmospheric components in determining weather and climate.

The hygrometric field corresponds to the presence and concentration of water in the air in the three physical states: solid, liquid and gas. This field also involves the transformation phases of the physical states, which are responsible for absorbing and releasing large amounts of energy.

Among the transformation processes, evaporation and condensation stand out mainly because of the way they define hygrometric behaviour. Evaporation is of significant importance because it is from evaporation that the atmosphere receives moisture from the earth's surface in bare soil, water surfaces and transpiration from plants and animals.

To differentiate between these two processes, Ayoade (2007) says that evaporation is the process by which moisture, in its liquid or solid form, passes into gaseous form. While the first term is used to describe the loss of water from water surfaces or bare soil, the second is used to describe the loss of water from surfaces with vegetation or living tissue.

These two processes are essential for the formation of air humidity, which refers to the presence of water vapour in the atmosphere. According to Torres and Machado (2011), air humidity results from the evaporation of water from land and water surfaces and from the transpiration of animals and plants, so it depends on heat and logically needs water to be evaporated.

From there, the terms that will classify and characterise the phases of air humidity will emerge. According to Mendonça (2007), terms such as vapour pressure, absolute humidity, specific humidity, mixing ratio and relative humidity are just variations in the way the presence of vapour is approached.

Of the terms presented, relative humidity is the best known, and certainly the most representative used to measure the presence of vapour, because it takes into account some local elements such as atmospheric pressure, temperature and air saturation in the area.

Relative humidity is defined by Ayoade (2003) as the ratio between the actual moisture content of an air sample, also called absolute humidity, and the amount of moisture that the same volume of air can hold at the same temperature and pressure when saturated, known as the dew point.

Relative humidity is closely linked to temperature, and this combination influences the rate of evaporation and transpiration, and consequently thermal sensation, since the balance between evapotranspiration and the amount of vapour present in the air determines the temperature felt by human skin, and as a result, human comfort, according to Sibbons (1962).

But it's worth remembering that absolute humidity, when it reaches the dew point, can saturate and then cause the moisture in the air to turn into water vapour and produce precipitation.

Water vapour is the origin of all forms of condensation and precipitation, so the amount of water vapour in a certain volume of air is an indication of the atmosphere's potential capacity to produce precipitation (AYOADE, 2007, p.128). This means that humidity not only directly influences the rainfall field, but also serves as the material component that feeds this process.

The rainfall field, which begins mainly with the process of cloud formation through the evaporation of water, has a direct impact on the occurrence of precipitation. However, according to Conceição (2008), the formation of clouds is not enough for precipitation to occur, because in addition to this process that occurs from condensation, there are several other processes for the formation of ice crystals and raindrops.

In order for the droplets to precipitate and reach the Earth's surface, before they evaporate and condense back into clouds, they must overcome barriers imposed by rising air currents. This is only possible if the clouds grow large enough through the process of boiling, where the molecules then aggregate and reach a size large enough to reach the surface.

Condensation of the vapour in the air results from a series of processes, starting with adiabatic cooling, with the slow and gradual rise of humid air, condensation and cloud generation and the saturation of the air and reaching the dew point temperature.

> According to Soares and Batista (2004), precipitation is the result of an advanced state of condensation. It occurs when the gravitational force overcomes the convection current and the moisture reaches the ground in liquid form (rain or drizzle) or solid form (hail, sleet and snow) (TORRES E MACHADO, 2011, p. 54).

Precipitation can also be an effective way of transporting much of the atmospheric pollution to the surface, because when the water that had been

evaporated into the air returns to its liquid state, its molecules adhere to other solid or gaseous particles in the air, bringing particulate matter, gases and other elements to the surface, leaving the air cleaner.

2.3. Climate and health

When pointing to meteorological elements and climatic factors as the cause of diseases, there is a risk of getting a false positive result, because many other issues may be involved in the occurrence of illnesses, which may not be linked to the climate. However, this does not mean that the climate does not have a direct or indirect influence on the occurrence of diseases.

> Finch and Trewartha (1954:64) stated that "since the human body is sensitive to factors other than air temperature, it is not a very accurate thermometer". "The body remains the most delicate recorder of complex climatological functions", in the words of Sorre (1955:23). Accept this fact and the fact that any associations made between environmental variables and human reactions are only approximations. They can, however, be more or less representative, depending on the number and quality of the climatic elements evoked (CONCEIÇÃO, 2008. p.52).

Just as the climate influences the lives of individuals, Nascimento (2011) states that the urban climate is considered to be the product of human action on the environment, essentially with regard to the process of urbanisation.

It is important to emphasise that illnesses generally occur due to the fragility of the organism, but aggravating factors contribute greatly to their incidence. In the case of meteorological factors, if we look at the most representative ones, i.e. those that always appear adjacent to illnesses, they can, depending on the other variables, be indicated as potential aggravating factors.

Ayoade (1986), for his part, emphasised that the influence of the climate on human health is both direct and indirect, whether in a harmful or beneficial way. For him, extremes of climate, especially temperature, disadvantage the body's ability to protect itself against illness.

> Fresh air (with a mild temperature, humidity and moderate radiation) has therapeutic properties. However, temperature can be the main trigger for some types of illness, more than any other climatic element, as is the case with infant mortality from respiratory illnesses and acute respiratory infections (SOUZA, 2008, p. 121).

Humidity has been identified as one of the main agents of respiratory illnesses, since a high incidence of cases occurs when relative humidity levels are very low. According to ABC da Saúde, with dry air, illnesses tend to worsen, since oxygen with low humidity enters the nose and lungs and can cause inflammation and excessive secretion production.

> Relative humidity (RH) can have direct and indirect effects on health and physical comfort. Both very high and very low humidity levels can affect health and well-being. These effects are minimised with humidity levels between 40% and 60%, which are considered adequate for preserving health (SMSSP, 2010, p. 1).

It is therefore considered that extremes of relative humidity, both maximum and minimum, can leave the human body vulnerable and expose it to various respiratory diseases.

> The direct effects of low RH on health stem from this condition on physiological processes and the indirect, more complex effects of its impact on pathogenic organisms or chemical substances. To prevent dry mucous membranes and maintain adequate mucus transport and ciliary activity, the RHU must remain above 40 per cent.

Another element of great significance is water vapour, as well as its quantity and distribution over time and space, since it is the origin of all forms of condensation and precipitation, according to Ayoade (2007).

The author also states that in a certain volume of air, humidity, in addition to producing precipitation, can also absorb both solar and terrestrial radiation and thus plays the role of thermal regulator in the Earth-atmosphere system, improving the body's thermal sensation.

Seasonal variations in rainfall can also have a direct influence on the occurrence of diseases, especially those of the respiratory system, because as well as having a direct relationship with relative humidity, rain plays a fundamental role in "cleaning the air" by depositing suspended material that adheres to the crystals that form the raindrops.

The great distance from water bodies alters the air temperature because of the basic differences in the thermal characteristics of continental and water surfaces (AYOADE, 2007).

In continental locations, the annual amplitude is greater and the delay between the maximum and minimum surface temperatures is only one month, while in coastal locations it can be up to two months.

This factor can be even more aggravating in the interior of the continent, because the shorter these periods are, the more difficult it is for the body to adapt to these ranges.

The prevailing winds influence the temperature, as they can transport or transmit heat or cold by advection. Seasonal variations in temperature are the result of the intensity of solar radiation received anywhere on the surface, in relation to the variation in the axis of the globe.

Winds are important in the absence of fronts and strong depressions, as they can disperse suspended elements and are fundamental for "air exchange". On the other hand, they can suspend materials deposited on the streets, which creates a layer of dust in the air. Their absence can have serious consequences for the population.

It is important to emphasise that, in addition to the atmospheric factors that can affect people's health, human actions can also cause reactions to atmospheric factors.

In this context, a debate arises that is part of discussions on public health, and is submerged by urban environmental issues and the search for sustainability and balance between the human-nature relationship. Environmental issues must involve, above all, the search for the core of quality of life, in other words, health and well-being.

It is therefore worth raising some of the environmental problems that arise from the way urban space is organised and through the intricate human relationships in cities, with a view to visualising sustainable possibilities.

Therefore, only those that are related to the atmospheric environment and air quality in the urban space and that can jeopardise people's respiratory system were analysed.

Air pollution has become one of the main urban environmental problems, especially in medium and large centres where there is a high concentration of people. It is almost always linked to human needs, such as the production of consumer goods and services, and the way in which resources are used to meet the demands of the globalised world.

Currently, one of the most important processes in this regard is the production and use of energy, because most human activities depend on energy, which generates enormous demand for this resource.

According to Goldemberg (2003), this production process plays the biggest role in atmospheric pollution, accounting for approximately 52 per cent of CO_2, as well as releasing other pollutants such as sulphur, nitrogen, particulate matter and ozone.

One of the main forms of pollution from energy use is through the burning of fossil fuels used for transport and solid fuels such as biomass and coal for heating and coercion.

According to COMPET (2011), a programme of the Ministry of Mines and Energy coordinated by Petrobras, the most obvious form of air pollution that we routinely encounter in urban centres, where there is a high concentration and movement of people and vehicles, is the waste produced by the combustion of diesel that drives the fleet of trucks and buses and which is released into the air we breathe.

Factories also play a major role in the emission of CO_2 into urban air, but as they are not always located in city centres or residential areas, they may be less important than vehicles, for example.

The emission of gases and particulate matter from cars and factories pollutes the atmosphere and retains heat, intensifying the greenhouse effect. This causes a number of problems for the population, because in addition to the health problems that individuals are exposed to, acid rain, thermal inversion and heat islands take on immense proportions, generating a chain of urban problems.

The disposal of rubbish and waste on the streets, mainly made up of small sediments such as sand and dust, can increase the amount of particles in the air, as can urban fires, which are a routine activity for burning rubbish and clearing vacant lots. This increases gases and fumes in the air, and the lack of green areas, or afforestation.

When proposing solutions for the urban environment, Loboda (2005) states that the presence of green areas can soften the thermal sensation, effectively filter solid particles suspended in the air and help maintain more humidity in the atmospheric environment. As well as these solutions, viable alternatives such as tarmacking and street cleaning can reduce the amount of dust blown by the wind, and as a result improve the air that is breathed in the urban space.

CHAPTER 3

METHODOLOGY

3.1. The method of approach

The development of knowledge through scientific research must be subject to a series of rules that should guide the work, and these are called scientific methodology. This is a set of approaches, techniques and processes used by science to formulate and solve problems of practical acquisition of knowledge in a systematic way.

Systematisation must go through several stages, where it must go through the necessary steps to order the activities to be carried out during the research, using reason as a tool, which can lead to scientific knowledge.

Thus, the method of approach that guided this work was the deductive method, which is based on the syllogism, where it starts from a generalisation to a particularised question, which accepts the arguments as true because they have already been validated by science.

This method allows the researcher to move from the known to the unknown. But it is limited, since the conclusion reached cannot go beyond the premises.

Alongside the inductive method, the analytical method was used along with the quantitative approach, which translates information into numbers to be classified and analysed using statistical techniques.

This made it possible to process and analyse the data collected, and also allowed for an in-depth reading of this information where other factors and elements were listed, resulting in a broader and more complex analysis.

3.2. Research stages

Systematised ordering in stages is essential for obtaining a satisfactory result in the time allotted. These stages followed a chronological sequence, with a defined timetable, in which some of them were worked on in parallel, as shown in the organisational chart in Figure 02 below.

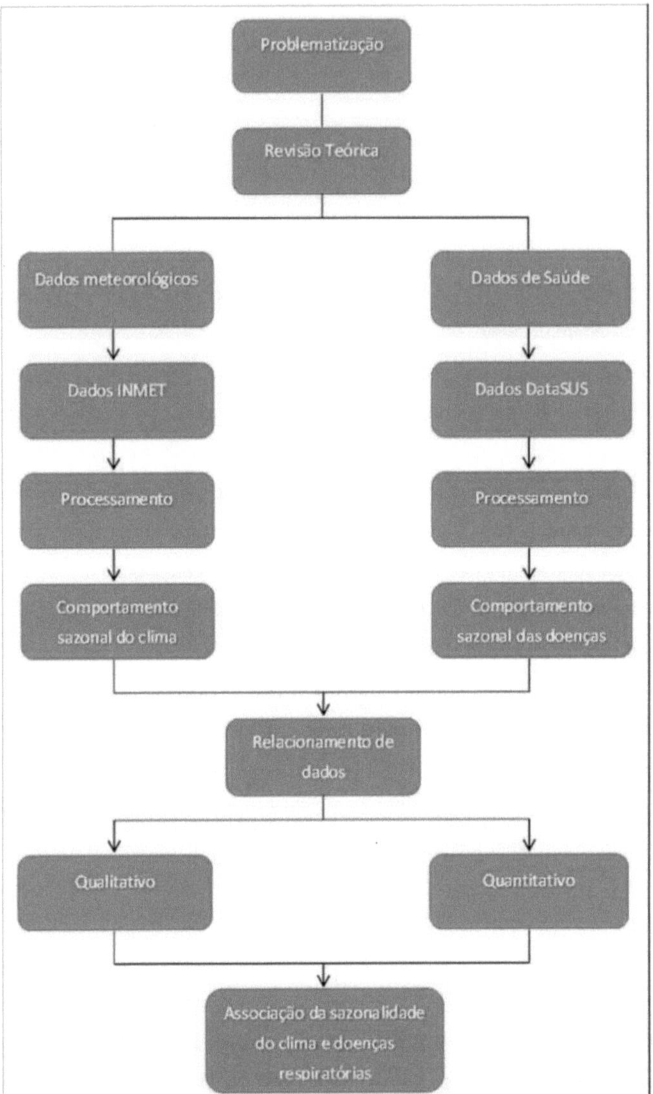

Figure 2. Organisational chart of the research stages

The first step was the desk work stage, which consisted of consolidating the research problem and the theoretical methodological framework, as well as a bibliographical survey of books, articles, dissertations, theses, websites and other theoretical sources.

In order to do this, it was necessary to make a preliminary enquiry, mainly regarding the resources and information available, to define the possibilities for carrying out this research and, from there, to determine the specific objectives, which also delimited the time frame of the research and the urban space of the Municipality of Barreiras as the spatial cut-off.

After problematising the issue, reading the bibliographies of authors who worked on the themes and concepts needed to understand the issue and the relationships discussed here served as a theoretical basis for constructing the literature review.

In the second stage, we searched for quantitative data records provided by the organisations and institutions responsible for carrying out research and gathering information, as well as disseminating it.

In this phase of the research, data was collected on morbidity[2] and hospital admissions, tabulated by the public health organisations provided by the Ministry of Health.

For the climate analysis, meteorological data was acquired from the Barreiras weather station, code A402, via the Brazilian Meteorological Institute (INMET). Statistical analysis was carried out in the third stage, by tabulating the data collected in the establishments and elsewhere, transcribing and interpreting the observations, drawing up graphs and tables, using Excel© Office 2010 as a support tool to better analyse and associate the variables.

Three types of data were used in this research: primary data, constructed by the researcher himself, such as tables, maps and graphs; secondary data, analysed by others; and the literature review, which is nothing more than the analysis of multiple trials (meta-analysis) that integrates the results.

The fourth stage, which is the final stage of the research, the results and discussions, is made up of the data analysis, which is divided into two parts: one descriptive, where the products of the statistical treatments appear, such as the values in averages and percentages, and the other qualitative, which shows in more depth through a comprehensive investigation derived from the quantitative analysis.

It is also at this stage that final considerations about the research are presented, where the results are discussed, the diagnosis of the research and proposals related to the topic are presented.

3.3. Data acquisition and processing

In order to carry out this work, it was necessary to investigate data sources that could provide two basic pieces of information for the municipality of Barreiras, one relating to climate records and the other to records of hospital admissions.

This was one of the most important stages in carrying out the research, since it could only be carried out using these data sources, given the complexity and cost of creating and maintaining a database of our own.

Initially, in the search for the municipality's health data, the Municipal Health Department was

---

[2] Morbidity is the quantification or rate of carriers of a given disease in relation to the total population studied, in a given place and at a given time, (BRASIL, 2005).

consulted, which did not have this data, but indicated a sub-secretariat called VIED (Epidemiological and Disease Surveillance). The 25ª DIRES (Regional Health Directorate of the State), which serves the municipalities of western Bahia, was also consulted, but no information on respiratory diseases was found.

The data concentrated at VIED is available on the DATASUS website, so there is no longer any need to travel to visit the municipal health centres.

All the information on hospital care provided by the SUS in the municipality is concentrated at VIED, where it is tabulated and made available through the TABWIN software in the Ministry of Health's national data system, called DATASUS, and is available for public access on the website www.datasus.gov.br, with the interface shown in Figure 3.

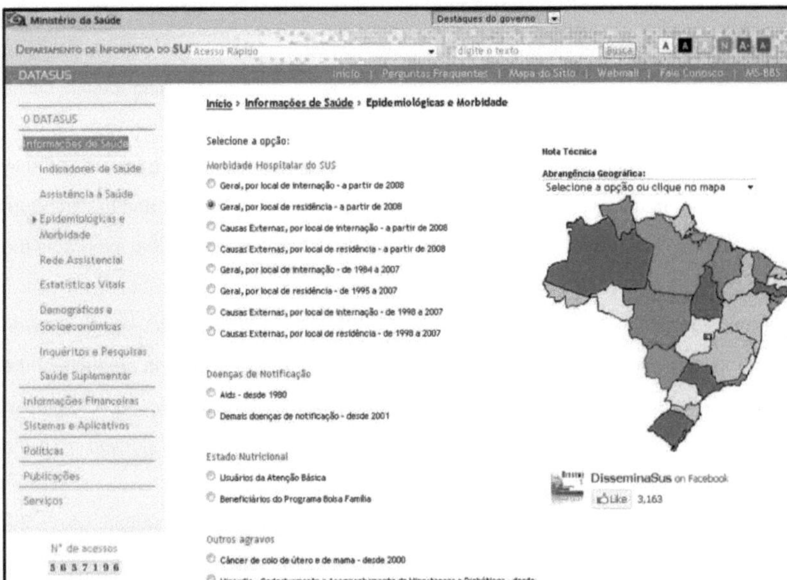

Figure 3 - Interface of the DATASUS health information page by epidemiology and morbidity.
Source: Brazil (2012)

For information on hospital morbidity, the options "general by place of residence from 2008 to 2010" and "general by place of residence from 1995 to 2007" were selected for monthly information. However, only data from 2000 to 2010 was grouped together.

The patient's place of residence is provided by the hospital where they were treated, and takes into account the city where they live, regardless of where they were treated. It should be noted that the information provided by DATASUS is based on hospital admission authorisations (AIH).

Another possibility consulted was by place of hospitalisation, however, this option only shows where the patient was treated, even if they come from another place, and has no relevance to the objectives of this research.By choosing the morbidity option and the State, the system, through the information sorting page, as shown in Figure 4, also allows you to filter the database by selecting a set of variables of interest.

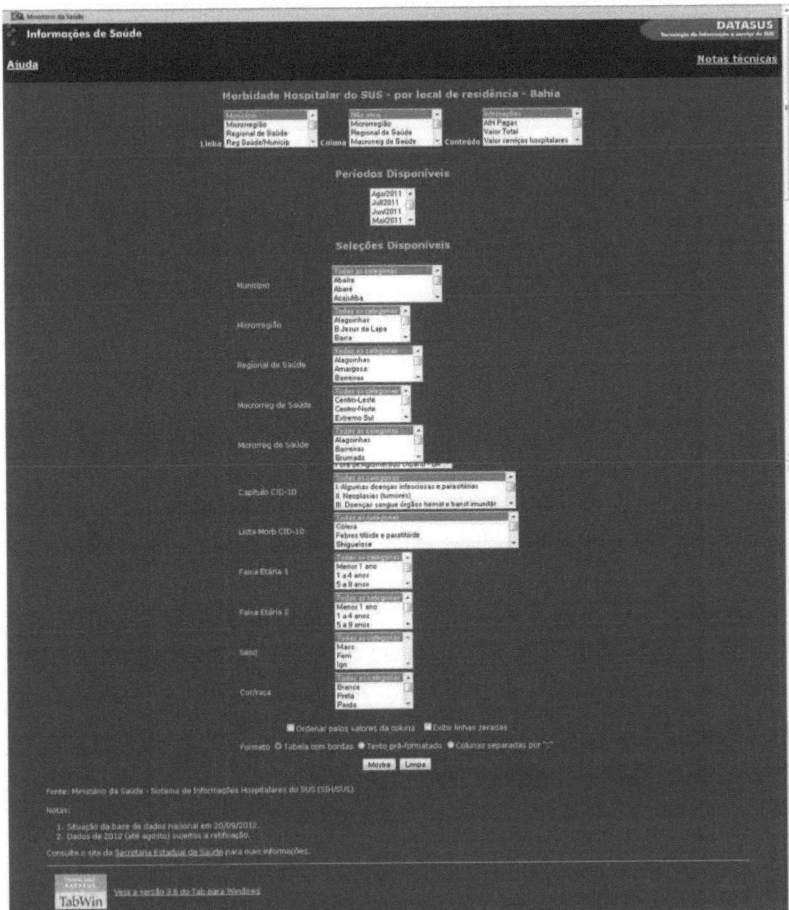

Figure 4 - DATASUS information screening page. Source: Brazil (2012)

The system has more than 50 variables, which enables personalised searches and the use of different analysis parameters. It thus provides a range of thematic data, such as hospital morbidity, mortality, vital statistics and tools to group or disaggregate information.

From there, health information on epidemiology and hospital morbidity by place of residence was consulted, with people living in the municipality of Barreiras, and by selecting the ICD-10 (10th International Classification of Diseases) Chapter X was selected, which covers diseases of the respiratory system as shown in Table 1.

Table *1.* Code and name of ICD-10 Chapter X diseases

ICD code	Chapter X ICD-10
165-179	**Diseases of the respiratory system**
165	Acute pharyngitis and acute tonsillitis
166	Acute laryngitis and tracheitis
167	Other acute upper airway infections
168	Influenza
169	Pneumonia
170	Acute bronchitis and acute bronchiolitis
171	Chronic sinusitis
172	Other diseases of the nose and sinuses
173	Chronic diseases of the tonsils and adenoids
174	Other upper respiratory tract diseases
175	Bronchitis, emphysema and other chronic obstructive pulmonary diseases
176	Asthma
177	Bronchiectasis
178	Pneumoconiosis
179	Other diseases of the respiratory system

Source: Brazil (2012)

In order to standardise classifications, according to Brasil (2012), ICD-10 was created to standardise and catalogue diseases and health-related problems, using the International Classification of Diseases, established by the WHO, as a reference.

Data was also collected on AIH admissions due to death and by age group. For cases of death, the system takes into account the number of hospitalisations that were discharged due to death, otherwise they are only quantified as hospital interaction.

The age group is classified by DATASUS into two categories: the first groups ages 1 to 19 every 5 years, and the second groups ages 20 to 79 every 10 years. The second category groups ages between 1 and 79 every 5 years. The categories also include a group with children under 1 year old and another with people over 80, as shown in Table 2.

Table *2.* DATASUS age group classification categories

Patient age groups, in the following categories:	
Age group 1:	**Age group 2:**
Under 1 year old	Under 1 year old
1 to 4 years	1 to 4 years
5 to 9 years	5 to 9 years
10 to 14 years	10 to 14 years
15 to 19 years old	15 to 19 years old
20 to 29 years old	20 to 24 years old
	25 to 29 years old
30 to 39 years old	30 to 34 years old
	35 to 39 years old
40 to 49 years old	40 to 44 years old
	45 to 49 years old
50 to 59 years old	50 to 54 years old
	55 to 59 years old
	60 to 64 years old

60 to 69 years old	65 to 69 years old
70 to 79 years old	70 to 74 years old
	75 to 79 years old
80 years and over	80 years and over

Source: Brazil (2012)

Age group 1 proved to be more appropriate for the research, which has a wider age range for adults and only emphasises age groups with a greater predisposition to illness, in the case of children.

The graphs were produced in Excel© from the monthly totals and averages of hospitalisations and deaths, as well as graphs by age group, highlighting the groups with the most hospitalisations.

There are numerous limitations to this data, which has led to restrictions in the possibilities for analysing the information. One example is the spatialisation of the data, since there is no information on patients' addresses and the place of care is not always in the same neighbourhood or sector of residence.

It is therefore not possible to specialise the data by zones, sectors or neighbourhoods, nor to distinguish between urban and rural patients, as most rural patients are treated in the city.

The meteorological data used in this research was recorded at the automatic station and the conventional station in the city of Barreiras, located at UTM latitudes E 500000.0 m and N 8645801.60 m, in Figure 05, available from the INMET (National Institute of Meteorology) database, via the website www.inmet.gov.br in the options under the Observations icon, Figure 06.

Figura 5. Barreiras Climatological Station. Source: INMET (2012)

The conventional station in Barreiras was opened on 1 January 1924 and the automatic station on 22 December 2001, with daily data for all those years. However, observations are made available up to 90 days prior to the date of access, meaning that to obtain meteorological data from these stations, a search must be made every three months at most on the INMET website, as shown in Figure 6.

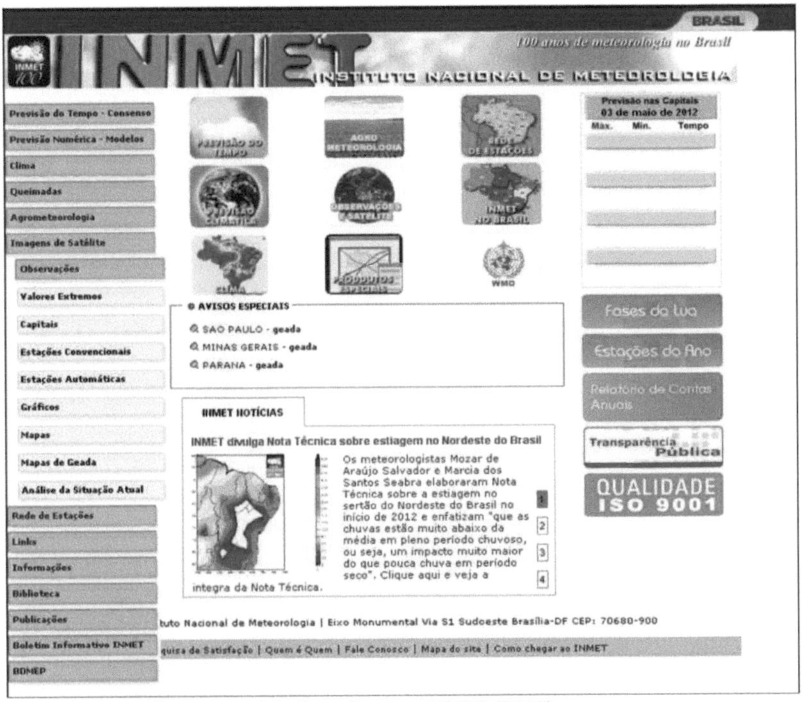

Figure 6 - Inmet homepage interface. Source: INMET (2012)

The only way to retrieve data from more than three months ago is to request INMET to evaluate it and release the data.

Daily data was collected from the conventional station at the three standard times observed in Brazil, i.e. 9 a.m. (12 UTC; Coordinated Universal Time), 3 p.m. (18 UTC), and 9 p.m. (0 UTC) every 90 days, from May 2011 to May 2012, for temperature, relative humidity, wind and precipitation.

For the same period, data was collected from the automatic stations, but the daily values for 24 hours starting at 0 h UTC (i.e. 9 p.m. the previous day) and ending at 23 h UTC (8 p.m.).

In addition to these collections, work was carried out to find data for the 10 years of coverage of meteorological behaviour from 2000 onwards. Thus, INMET was asked by fax to provide monthly meteorological data for the period from January 2000 to December 2010,
requiring the following parameters: air temperature (maximum, minimum and average), relative humidity, rainfall, wind direction and speed.

The data was received via email and, as it had already been processed into monthly averages, only the annual averages, annual maximums and minimums, percentages for the dry and rainy periods, and the percentage increase were calculated, where, based on this information, graphs were drawn up in an Excel spreadsheet to associate and analyse them.Heat spot data from 2000 to 2010 was also needed to associate with the health data. To do this, we consulted the INPE fires database on the website http://www.dpi.inpe.br/proarco/bdqueimadas, in Figure 7, which shows the hotspots identified by the NOAA, GOES, AQUA, METEOSAT and TERRA satellites.

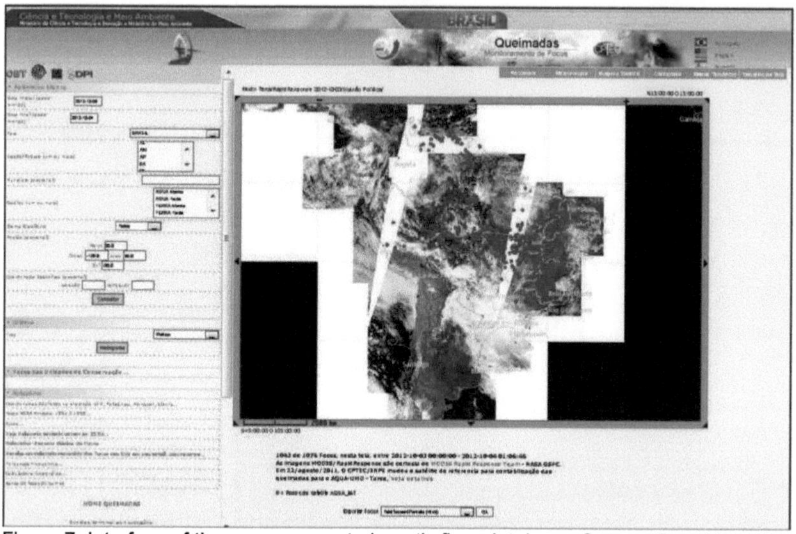

Figure 7. Interface of the access page to Inmet's fires database. Source: Brasil (2012)

The hotspots were then acquired in shapefile format, for each year, and 5 km buffers were created using ArcGis 9.3 *software*, 10 km, 15 km and 20 km around the urban area, using the *Proximity Buffer* tool. The *clip* processing of the *Extract* tool and the *Erase* and *Spatial Join* processing of the *Overlay* tool were also used.

The *Join Count* column of the attributes table shows the number of heat spots recorded for each year within the urban area, up to 5 km, from 5 to 10 km, from 10 to 15 km and from 15 to 20 km from the urban area.

Photographic records were also taken of the urban atmosphere of Barreiras, on a weekly and monthly time scale, from two observation points that provide a panoramic view of the city. One is located at Serra do Mimo, at UTM coordinates E 501467.14 m; N 8656646.70 m, and the other at Serra da Bandeira, located at UTM coordinates E 500500.67 m and N 8662034.05 m.

The photographs were processed in Microsoft Office 2010's Picture Manager, and were processed for horizontal panoramic viewing in Arcsoft's Panorama Maker software.

3.4. Parameters for analysing and associating data

Analysing health data as well as climate data can involve very broad issues and it is not possible to cover everything related to this topic in a single study. The use of parameters to limit analyses allows approaches and associations to be more correlated, even if they involve many variables.

Thus, in order to address the association between meteorological behaviour and hospital admissions for respiratory diseases, some qualitative parameters were used, through the analysis of graphs and trend lines and the observation of photographic records of the urban atmosphere, and quantitative parameters, using descriptive statistics, whose parameters were defined by means, amplitudes, percentages, maximums and minimums, graphs of deviations and variances.

CHAPTER 4

RESULTS

4.1. Barreiras' climatic seasonality

In this first subchapter of the results, meteorological records of rainfall, relative humidity (RH), temperature and temperature range were analysed with a view to analysing the existence of seasonality in the climatic behaviour of the city of Barreiras in relation to these elements.

Before presenting the results of the association between meteorological behaviour and the occurrence of respiratory tract diseases (RDS), it is necessary to characterise the city's climate, because only from this point will it be possible to verify the relationship between meteorological and epidemiological variables.

Conceição (2003) identified that there is a common use among geographers and doctors in this area of knowledge, in relation to the use of monthly values to explain the causality between climatic elements and health. In this sense, the monthly values of climate elements recorded at the automatic and conventional meteorological stations in Barreiras will serve as a basis for the association between climate and health.

The meteorological elements dealt with here follow the perspective of Corrêa (2000), who identifies in his research on climate, environment and human health that thermal conditions, air humidity, precipitation and dispersion (winds and pollution) are more important in the causality approach.

The climatic characterisation was based on the analysis of two historical series of climate data, the first referring to data from the Climatological Normal drawn up by INMET for the period 1961 to 1990, and the second, also drawn up from INMET data, which presents meteorological records from January 2000 to December 2010.

It was necessary to work with the second historical series because the observations of hospital admissions also cover this period, which makes it easier to associate them.

As discussed in the theoretical framework, Ramos (2009) states that the WMO establishes the climatic characterisation of an area by defining a meteorological observation over a long period of time, with a recommendation of at least 30 years of observation.

The averages observed at the station from January 2000 to December 2010 follow the instructions of the WMO, which, according to the aforementioned author, states that if a Climatological Normal is not available for a period, provisional normals can be calculated, with a minimum of 10 years of observation.

The two series of meteorological observations totalled 41 years of observation, but were worked on separately because there was a gap of unavailable data of 10 years between them (from 1990 to 1999).

Souza (2005) states that in some tropical regions, especially the Cerrado, due to seasonal variations, the hydrological year is defined in two half-yearly seasons, the dry and wet half-years. Based on this statement, in analysing the Climatological Normal, the monthly rainfall averages were divided into two hydrological periods, shown in Graph 1, using an annual average line, calculated from the monthly averages of the twelve months of the year.

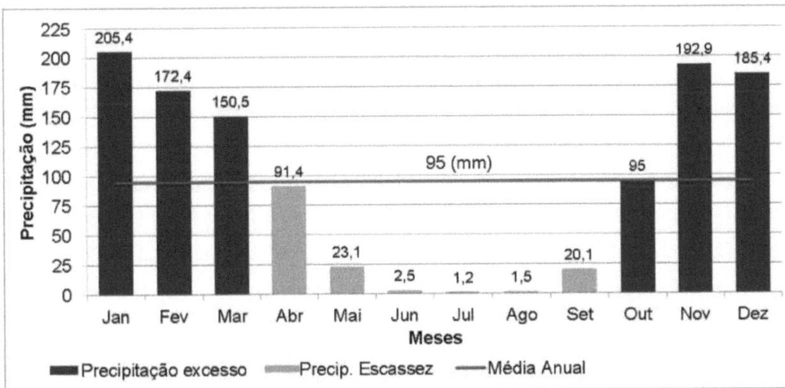

Graph 1. Average monthly rainfall in Barreiras (1961 to 1990). Source: INMET, (2009)

The months from October to March are therefore defined as the hydrologically humid semester, since the average monthly rainfall was higher than the average monthly rainfall of 95 mm. The hydrologically dry half of the year, which runs from April to September, includes the months whose average rainfall was lower than the monthly average mentioned above.

Graph 1 also shows that rainfall in the six months with the highest rainfall totalled 998 mm, and only 142 mm for the other six months with the lowest rainfall. This means that 88% of the total rainfall is concentrated between the months of October and April, while 12% is distributed unevenly between the months of May and September.

The months of most intense water scarcity, usually June, July, August and part of September, have average rainfall volumes of less than 5 mm per month and produce very dry weather conditions in the atmospheric environment of this region.

It was found that the months with excess rainfall in the study area, i.e. above the monthly average, between October and March, especially in summer, are the wettest period of the year and coincide with the periods when the humid air masses mEc and mTa act with greater intensity, as shown in Figure 8A.

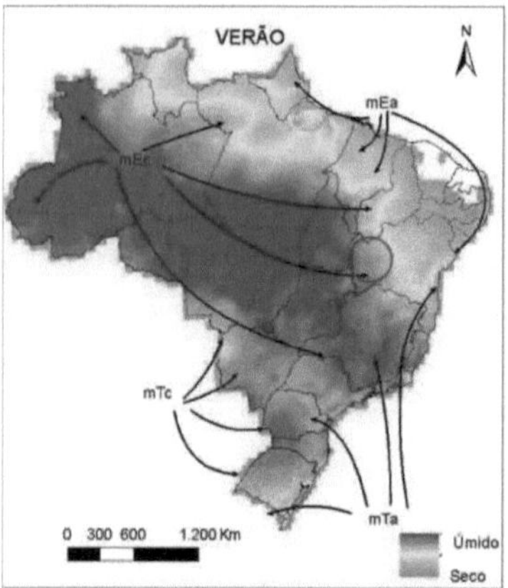

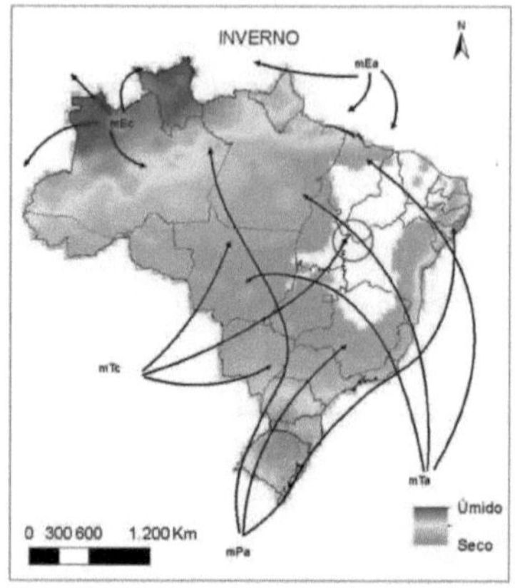

Figure 8A. Air masses in Brazil during the summer. Source: Mendonça (2007). Adapted from Porto (2012)
Figure 8B. Winter air masses in Brazil. Source: Mendonça (2007). Adapted from Porto (2012)

Figure 8B, on the other hand, shows that the months from April to September have a rainfall below the average monthly value, which usually occurs in winter, making it the period with the lowest volume of precipitation in western Bahia, with little or almost no rainfall.

In winter, the main masses that bring moisture to these areas (mEc and mTa) recede, and the hot, dry air masses (mTc) and the cold, dry air mass (mPa) advance across the continent and cause periods of water scarcity in the Brazilian Centre-West and much of western Bahia.

Due to the retreat of the mEc, the mTa is advancing towards Central Brazil. However, as it is a thin layer mass, it loses moisture in the coastal areas, ceasing to be humid and cold, and becoming dry and hot, which significantly reduces the humidity in the centre of the continent.

However, as identified by Mendonça (2007), in addition to dry and hot air masses, some geographical factors, such as continentality and relief, can significantly contribute to intensifying these dry periods.

When working with areas of the Cerrado, Souza (2005) identified, based on absolute rainfall data, that the region has some characteristics similar to those of semi-arid or desert climates, such as low RH, with daily absolute percentages of less than 10%.

The period of greatest water scarcity is also characterised by significant daily temperature variations, according to observations by INMET (2012), with average monthly variations of up to 20X3, which can reach much higher daily values of more than 25X3.

Graph 2 shows that the lowest amplitudes occur from October to April, during the periods of greatest rainfall, because humidity acts as a temperature regulator, since water absorbs and loses heat slowly. However, the amplitude can be reduced by other factors, such as cloudiness and daily sunshine, which can be associated with precipitation.

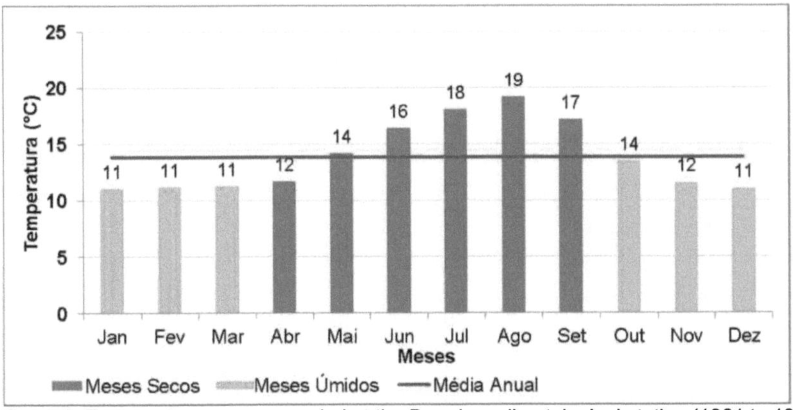

Graph 2. Temperature range recorded at the Barreiras climatological station (1961 to 1990).
Source: INMET (2009)

Cloudiness is an extremely important factor in this process, as it reduces the penetration of the sun's rays to the surface, which lowers the surface temperature. In addition, they act as a "lid", which reduces the exchange of heat between the surface and the atmosphere, thus maintaining the balance of atmospheric temperature, which inhibits very low or high temperatures, which reduces the daily variation during this period.

Graph 2 shows that the bars with values above the average monthly amplitude of 130 coincide with the months of greatest water scarcity, from April to September, when the variation between maximum and minimum temperatures is significant. In these months, rainfall is scarce and the RH is very low, which favours greater acceleration in the processes of heat loss and gain (DEFFUNE, 1994).

Daily insolation stands out in this process, since the reduction in cloudiness causes an increase in the amount of solar radiation, which also causes an increase in surface temperature. The decrease in cloudiness is related to the low humidity of the air, since the lack of water vapour in the atmospheric environment makes the condensation process for cloud formation unfeasible, as already discussed by Ayoade (2007), and this causes the clouds that arrive in the region to disperse quickly.Thus, the low humidity of the air causes a rapid exchange of energy between the air and the surface, which results in sudden fluctuations in atmospheric temperature, as shown in Graph 3.

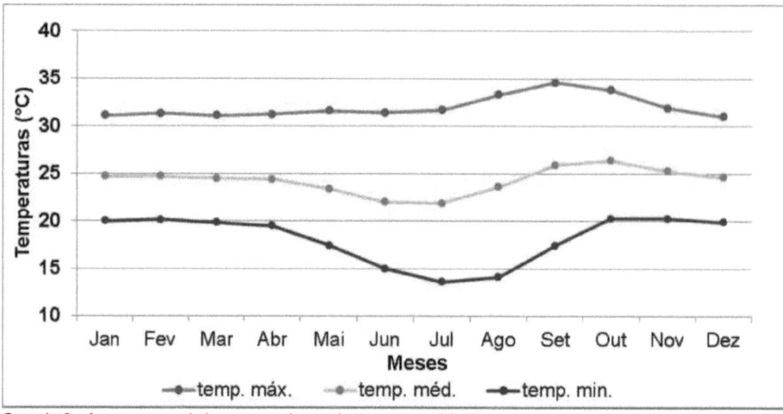

Graph 3. Average, minimum and maximum monthly temperatures recorded at the Barreiras climatological station (1961 to 1990). Source: INMET (2009)

The results observed in relation to the increase in thermal amplitudes in the INMET records (1961 to 1990) show that this is mainly due to the variation in monthly minimum temperatures, since there is a variation of more than 60 between the highest and lowest monthly minimum temperatures, while the variation between the highest maximum temperature and the lowest maximum temperature is only 3.50.

This means that the average maximum monthly temperatures do not fluctuate much throughout the year, except for the months of August, September and October, while the minimum temperatures fluctuate the most, especially in the months of April to September, causing a downward oscillation in the average minimum temperature line.

It can also be seen that the months of July, August and September, as well as having the highest temperature ranges, above 170, as seen in Graph 2, also have the highest maximum temperatures, up to 350 in September, and the lowest minimum temperatures, below 150 in July.

The relative humidity values shown in Graph 4 reveal that there is a significant difference in the percentages between the months considered hydrologically dry and those considered hydrologically humid. This observation can be seen by analysing the monthly values and the monthly average RH, with an average of 68%, where the months from November to May are above average, while from June to October the percentages are well below average.

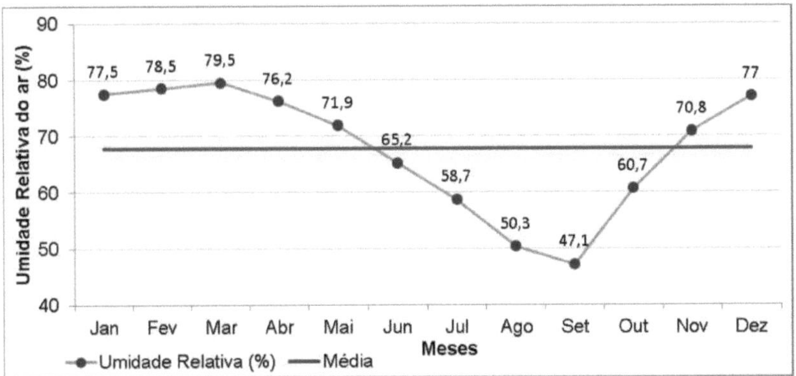

Graph 4. Average relative humidity recorded in the city of Barreiras (1961 to 1990). Source: INMET (2009)

The relative humidity averages between November and May show high percentages, with average values between 70% and 80%, and absolute values above 90%, close to air saturation.

The set of processes that influence humidity is fundamental to "cleaning the air" in cities, because water vapour adheres to particulate matter, which becomes heavier and is carried to the surface by gravitational transport.

In the months of June to October, humidity levels reach percentages below average, close to 56.4%, and can show absolute values that reach levels below 10%. These extreme values are recorded mainly around 3pm, as can be seen in Graph 5.

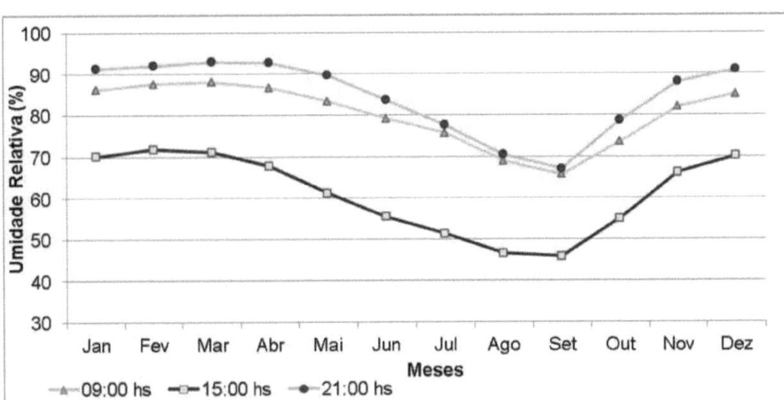

Graph 5. Average hourly relative humidity recorded in the city of Barreiras (1961 to 1990). Source: INMET (2009)

Graph 7 also shows the monthly hourly humidity averages at three different times, recorded at 9am, 3pm and 9pm, which reflect the daily behaviour of the RH. It can be seen that the lowest average RH recorded during dry periods usually occurs around 3pm, with percentages below 45%, when radiation and, consequently, temperature, are at their highest.

This process occurs because the atmosphere, especially in urban environments, expands at the hottest times, causing the humidity to be distributed over a larger volume of air, which means that water vapour becomes scarcer as the air expands.

Ayoade (2007) states that the concentration of atmospheric air leads to an increase in relative humidity, as well as in the amount of particulate matter in the air. In the city of Barreiras, this factor is intensified by the slopes of the hills that surround the city.

This accentuates the concentration of the atmospheric layer, as shown in the topographic profile in Figure 9, which, as well as hindering atmospheric circulation, causes temperatures to rise in the lower areas.

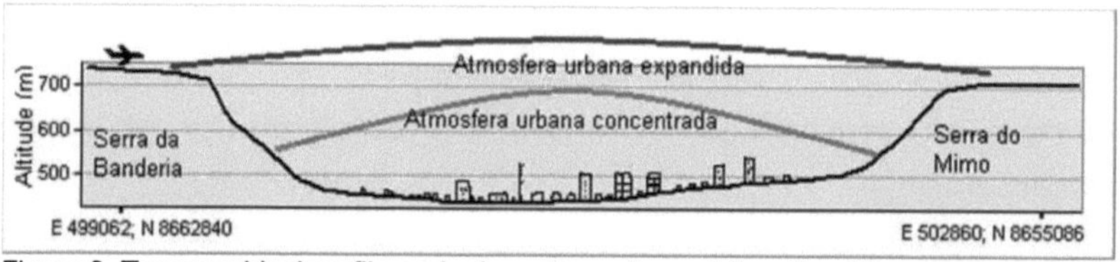

Figure 9. Topographical profile and urban atmosphere of the city of Barreiras

Thus, the definition of Barreiras' climatic behaviour using the Climatological Normal from 1961 to 1990 made it possible to identify two seasonal seasons in the city of Barreiras: a rainy season, with low temperature variations and high RH, and a dry season, with high temperature variations and low RH.

4.1.1 Analysing seasonality from 2000 to 2010

Analysis of the meteorological records from the Barreiras station from 2000 to 2010, as well as the Climatological Normal from 1961 to 1990, showed the existence of climatic seasonality in two periods of the year, which also reveal the occurrence of some variations in climatic behaviour.

The eleven years of observation, as already mentioned, are of significant importance to the results of the research, since they can be associated with the same period in which the ARD data was obtained. They were worked on the basis of WMO recommendations, according to Ramos (2009), from which these values were defined as the provisional Climatological Normal.

The results observed from this data, through the meteorological records of the Barreiras automatic station (Code 83236), show that there has been a "slight" monthly shift in the behaviour of rainfall in Barreiras, compared to the data from the Climatological Normal (1961-1990). This is because the period with above-average rainfall, i.e. the hydrologically wet months, begins in November, rather than October as previously, and lasts until April, which can be seen in Graph 6.

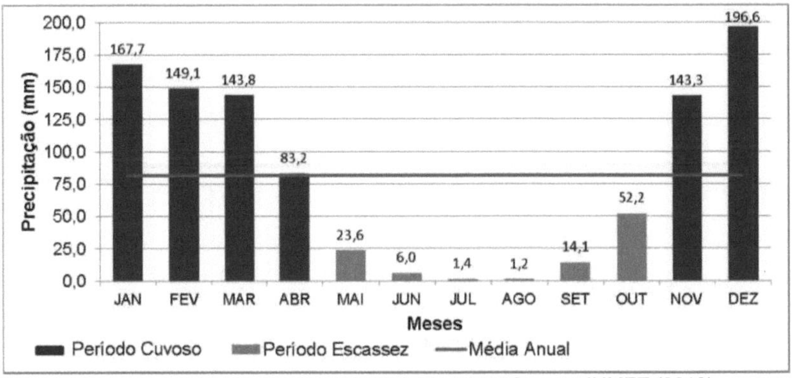

Graph 6. Monthly rainfall distribution from 2000 to 2010. Source: INMET (2012)

On the other hand, the hydrologically dry months, which used to start in April, started in May and lasted until October, with an average of 16.4 mm. This indicates that there may have been a significant change in the city's climatic behaviour, due to various hypotheses, including the growth of the urban sprawl.

In addition to the fact that there was a shift of one month in this period in relation to the data from the Climatological Normal, there was a decrease in the volume of rainfall, since the sum of the monthly averages for this period showed an average of only 98 mm, compared to 142 mm for the dry months of the normals, i.e. there was a 31 per cent drop in rainfall in the dry months.

Considering that the rainiest period also sees a decrease in rainfall, from an average of 1,000 mm per year to an average of 884 mm, a decrease of 12 per cent. This decrease in annual rainfall can have a number of impacts, especially on health, which can be very relevant in terms of worsening respiratory diseases.

However, there are two prerogatives to be considered: the first refers to the length of the series, which is considerably short for a climatic characterisation, and the second refers to the possible influence of the urban microclimate, due to the expansion of the urban sprawl, but this can only be confirmed through more in-depth research.The pattern observed in the monthly averages of relative humidity, shown in Graph 9, proves that it follows the behaviour of rainfall. This shows that rainfall and humidity are highly correlated, as an increase in RH favours the condensation of water vapour and the formation of clouds, from which precipitation can occur, which consequently also increases RH.

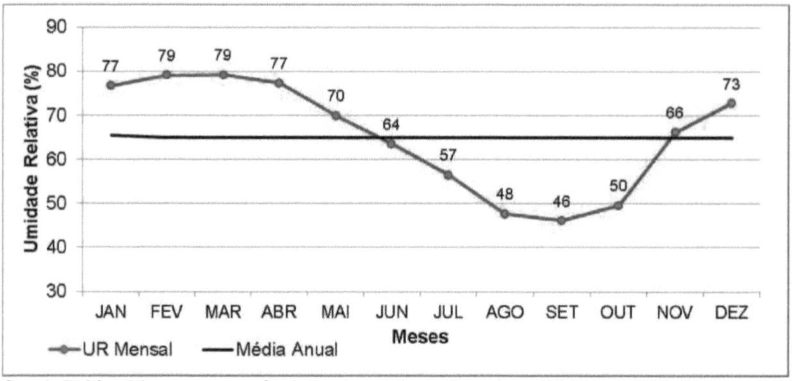

Graph 7. Monthly averages of relative humidity in Barreiras (2000 to 2010). Source: INMET (2012)

There is also a seasonal behaviour in RH, as shown in Graph 7, in which the months from November to May show percentages above the average monthly value of 65%. The months from June to October show values below 65%, which can be characterised as hydrologically dry months with low relative humidity.

However, the daily values show that the lowest percentages of humidity occur on the days with the highest temperature variations, as can be seen in Table 3, which shows the temperature and relative humidity values for the 5 driest days identified in 2011.

Table 3. Absolute values of the days with the lowest humidity in 2011

Date	Temperature fC)				RH (%)		
	Average	Maximum	Minimum	Amplitude	Average	Maximum	Minimum
10/09/2011	26	37	14	23	33	73	11
11/09/2011	27	37	16	21	27	55	10
12/09/2011	26	37	13	24	36	68	10
13/09/2011	27	36	15	21	30	65	11
14/09/2011	25	36	14	22	32	64	13

Source: INMET (2012)

These variations are exemplified in Graph 8, which shows the temperature meteogram in Barreiras in July, and Graph 9, the relative humidity, both available from the CPTEC (Centre for Weather Forecasting and Climate Studies).

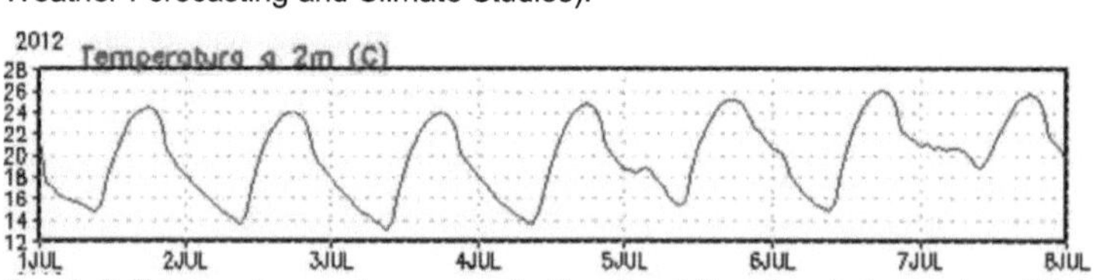

Graph 8. Temperature meteorogram for the city of Barreiras between 1 and 8 July 2012. Source: CPTEC (2012)

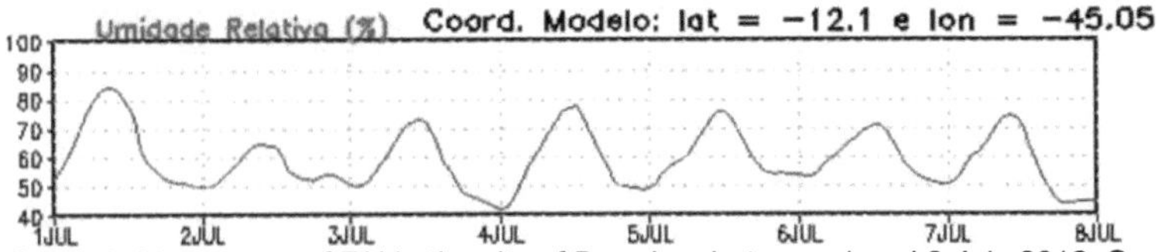

Graph 9. Meteogram of RH in the city of Barreiras between 1 and 8 July 2012. Source: CPTEC (2012)

The results observed in the daily temperature variation, using 2011 data from INMET, can be partly explained by the behaviour of the RH, because, according to the association between Graph 8 and Graph 9, the relationship between temperature and RH is inversely proportional, which becomes even more evident when there is a shortage of precipitation, since there is greater variability in the percentage of humidity, and consequently in temperature.

In the temperature range data from 2000 to 2010, Graph 10 shows a monthly average of 13.513, very close to the average observed in the Climatological Normal from 1961 to 1990, which was 13.8X3. Thus, there was a decrease in the monthly temperature range, of 0.513 on average, and more than 1*C from July to September.

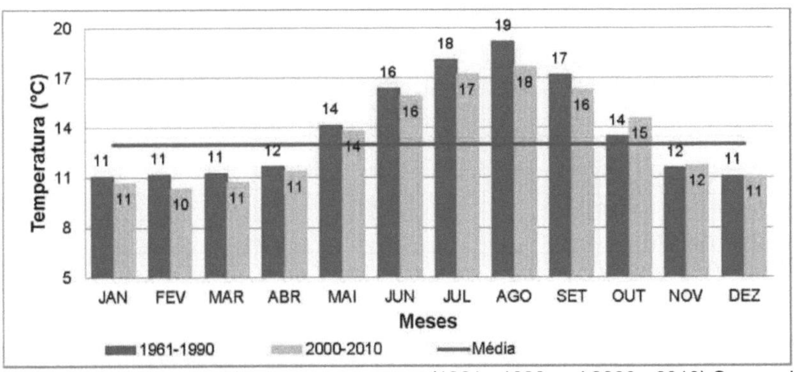

Graph 10. Average monthly temperature range (1961 - 1990 and 2000 - 2010) Source: INMET (2012)

The decrease in amplitude can be explained by the increase in minimum temperatures, as well as maximum temperatures, the latter being less expressive, which may also be an effect of the increase in urban sprawl. The opposite process occurred for the month of October, which had an average amplitude of 1513, with an increase of 1.413 compared to the Climatological Normal (1961-1990).

In the months of July and August, it was found that the significant amplitudes may be associated with the RH, since the lower the relative percentages of humidity, the greater the amplitudes, since in these conditions, temperatures increase with the incidence of solar radiation, and decrease considerably at night, when there is no solar radiation and little humidity.

Therefore, from the graphs presented, it can be seen that there is a correlation between the meteorological factors of rainfall, relative humidity, temperature and temperature range, which allow this climate pattern to be classified as a Tropical climate, with the presence of two well-defined seasons.

Even with two well-defined seasons in the study area, there are still particularities in each season, where months or days with extreme values, for example, greater meteorological variations, can be aggravating factors for the occurrence of respiratory diseases.

The analyses of the Climatological Normal (1961-1990), as well as the meteorological records (2000-2010), resulted in Table 4, which shows the months and average values of the seasonal stations in the city of Barreiras.

Table 4. Seasonal periods of meteorological behaviour in Barreiras

Year	**1961-1990**		**2000-2010**	
Seasonal period	**Rainy**	Dry	**Rainy**	**Dry**
Months	January	April	January	May
	February	May	February	June
	March	June	March	July
	October	July	April	August
	November	August	November	September
	December	September	December	October
Precipitation (mm)	1.000,2	139,8	883,7	98,4
Average temperature range (O)	12	16	11	16
Average maximum temperature (O)	31	32	32	34
Average minimum temperatureCC)	20	16	21	18
UR Average (%)	74	61,5	65	56

Source: INMET (2012)

The classification of Barreiras' climatic seasonality into a hydrologically dry period and a hydrologically wet (rainy) period makes it clear that the city's seasonality is mainly due to rainfall.

The hydrologically humid or rainy period is characterised by significant rainfall, with an average of 147mm per month, where relative humidity is high, on average 75%, and temperatures are relatively high, with an average of 25X3.

However, during this period, the temperature ranges have lower monthly averages, approximately 11X3, and comprise the months of November to April, as shown in Table 2. During this season, the body feels less fatigued because the air is more humid and the temperatures are milder.

According to INMET data from 2000 to 2010, the seasonal dry period was characterised by low rainfall, with an average of 98.4 mm over the whole period (six months), which had low percentages of RH in the air, and prolonged droughts of up to seven months, with high maximum temperatures and short minimum temperatures, which causes an increase in the temperature range. This period runs from May to October, but in August and September there are extreme shortages, with humidity below 10% and a temperature range above 20 C.<

It should be emphasised that even during periods of water scarcity, it is extreme weather situations that can be linked to an increase in hospitalisations for diseases related to the atmospheric environment, especially respiratory diseases, as discussed in the following chapter.

4.2. Occurrence of hospitalisations due to respiratory problems

4.2.1. Population growth in recent decades

The city of Barreiras is an important centre in the west of Bahia. It is the largest municipality in terms of population in the region and polarises the goods and services sectors. Its population growth has accelerated over the last three decades, according to Table 5.

Table 5. Growth of population in the city of Barreiras (1970-2010)

Population	1970	1980	1991	2000	2010
Total Population	20.864	41.462	92.640	131.849	137.428
Male	10.064	20.484	46.778	66.177	67.892
Female	10.800	20.978	45.862	65.672	69.536
Urban	9.760	30.055	70.870	115.784	123.734
Countryside	11.104	11.407	21.770	16.065	13.694
Urbanisation rate	46,80%	72,50%	76,50%	87,80%	90,04%

Source: GEOSNIC (2012)

Population growth was mainly due to increased urbanisation, as shown in Graph 11. As well as the expansion of the urban sprawl, this process led to a decrease in the rural population, driven above all by the displacement of this population to the city, and by the growth of agribusiness with the formation of large estates and a decrease in family farming.

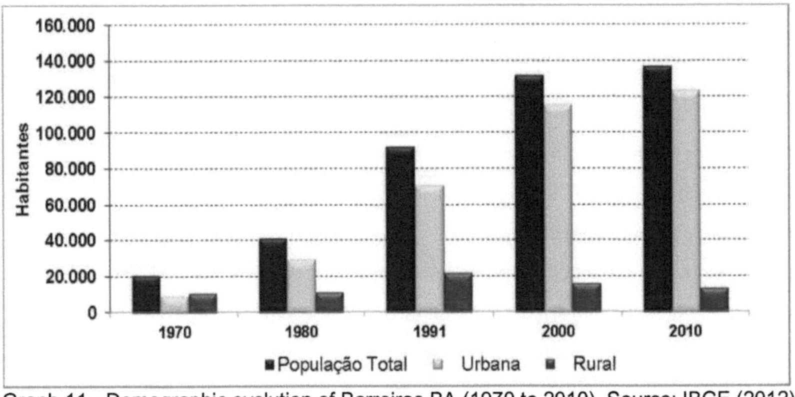

Graph 11 - Demographic evolution of Barreiras BA (1970 to 2010). Source: IBGE (2012)

Population growth has led to the expansion of the urban sprawl, as shown in Figure 10, which has resulted in an increase in urban problems, such as those related to public health and urban atmospheric problems.

Urban planning was unable to keep up with the speed of population growth, which did not prioritise basic infrastructure services during this period, such as sewage disposal, traffic, rainwater drainage, street lighting, public transport, green areas and communal spaces.

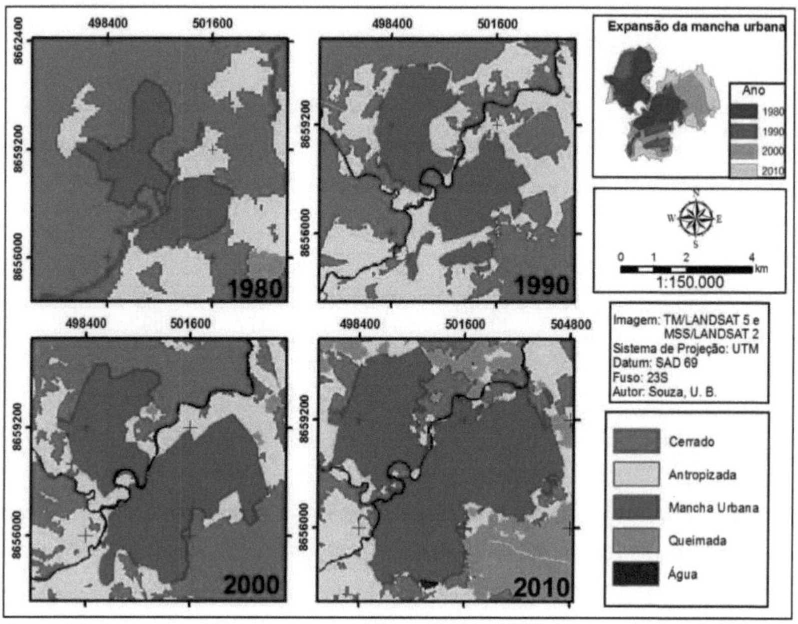

Figure 10: Expansion of the urban sprawl from 1980 to 2010. Source: Souza (2012)

Urban planning was unable to keep up with the speed of population growth, which during this period did not prioritise basic infrastructure services such as sewage, traffic, rainwater drainage, street lighting, public transport, green areas and communal spaces.

Urban disorder can have a direct effect on the quality of life of the urban population and have a significant impact on public health, since these factors in the urban environment are part of health promotion strategies to prevent disease.

The age distribution of the population in Barreiras, as shown in Graphs 12A and 12B, shows that in the last decade there has been a decrease in the birth and death rate in the municipality. This is because, according to the IBGE Demographic Census (2010), the percentage of children aged 0 to 9 in 2000 was 26 per cent, with 30,117 people, and fell to 17 per cent in 2010, with 23,777 people, i.e. a percentage reduction of 21 per cent, which caused the base of the demographic pyramid in 2010 to narrow.

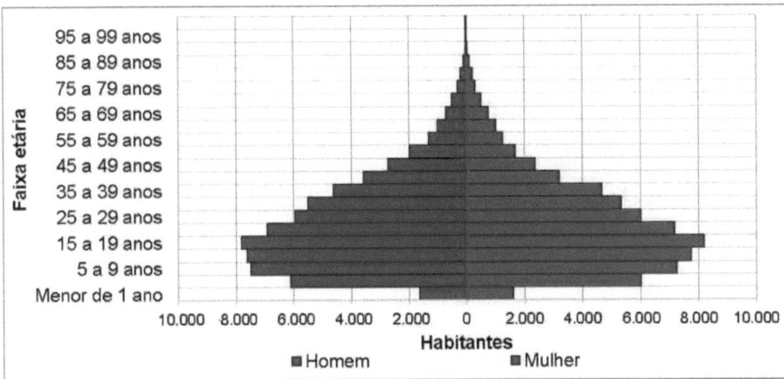

Graph 12A. Resident population age groups from the 2000 Census Source: IBGE (2011)

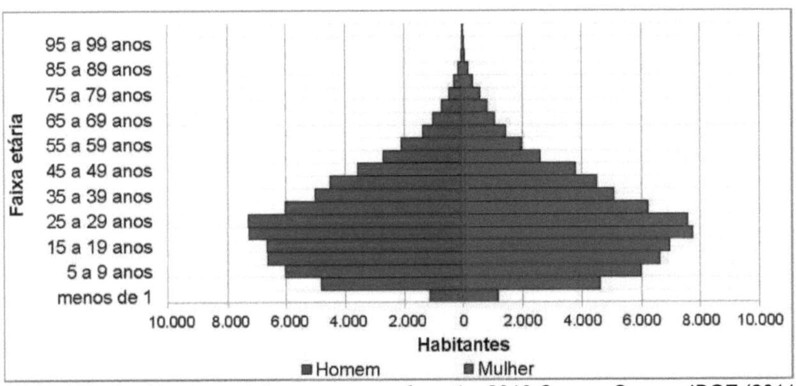

Graph 12B. Resident population age groups from the 2010 Census Source: IBGE (2011)

The fall in the birth rate may also have fuelled the decrease in the population under 18, as it fell from 55,026 in 2000 to 45,192 in 2010, a reduction of 18%.

The reduction in mortality was confirmed by the increase in the percentage of elderly people over 60, who represented 5 per cent of the total population of Barreiras in 2000, with 5,941 inhabitants, and went on to represent 6 per cent in 2010, with 8,727 people, an increase of 46.8 per cent according to IBGE (2010), which was possibly driven by the increase in life expectancy in Brazil.

The expansion of the trunk of the 2010 pyramid in relation to the 2000 pyramid is reflected in the growth of the adult population, which is between 18 and 60 years old, from 70,883 to 83,480 people, an increase of 19 per cent, which indicates that, in addition to the reduction in mortality, there has been an "ageing" of the population in general.

4.2.2 Classification of the main causes of hospital morbidity in Barreiras

Diseases affecting the respiratory system have stood out in Barreiras as one of the most serious problems related to the health of the population, due to the number of hospitalisations over the last decade, according to DATASUS (2012), with an average of 1,600 hospitalisations per year, as shown in Table 6.

Table 6 SUS hospital morbidity in Barreiras from 2000 to 2010

		Hospitalisation by hospital morbidity from 2000 to 2010 - Barreiras-BA												
Position	**ICD-10 Chapter**	**2000**	**2001**	**2002**	**2003**	**2004**	**2005**	**2006**	**2007**	**2008**	**2009**	**2010**	Total	Percentage
1	XV. Pregnancy, childbirth and the puerperium	4.436	3.861	4.097	3.819	3.690	4.143	4.308	4.235	2.811	2.858	2.854	**41.112**	31%
2	X. Diseases of the respiratory system	3.193	3.072	3.015	2.448	1.715	1.661	1.527	1.784	1.592	1.782	1.536	**23.325**	18%
3	I. Infectious and parasitic diseases	1.430	1.870	1.578	1.160	867	535	579	1.098	1.780	2.498	2.227	**15.622**	12%
4	XVI. Conditions originating in the prenatal period	2.433	872	1.330	2.025	1.256	1.562	1.435	763	389	347	140	**12.552**	10%
5	XIX. Poisoning and other injuries	694	446	846	1.006	1.097	928	1.108	1.231	639	666	797	**9.458**	7%
6	XI. Diseases of the digestive system	819	678	754	725	625	507	843	867	748	953	900	**8.419**	6%
7	IX. Diseases of the circulatory system	903	720	737	876	689	632	729	670	610	661	752	**7.979**	6%
8	XIV. Diseases of the genitourinary system	542	448	417	554	607	402	542	509	578	672	716	**5.987**	5%
9	XIII. musculoskeletal and connective tissue diseases	440	342	504	429	355	245	325	317	243	232	210	**3.642**	3%
10	XVIII.Symptoms or signs and abnormal clinical and laboratory findings	595	1.263	196	198	143	120	150	177	80	95	93	**3.110**	2%
11	Other chapters	670	656	1.010	1.096	953	794	1019	1.006	1.129	1.121	1213	10667	8%
Total		15.485	13.572	13.474	13.240	11.044	10.735	11.546	11.651	9.470	10.764	10.225	131.206	100%

Source: DATASUS (2012)

Diseases of the respiratory system (DAR) are also noteworthy, as they occupy the second position in the overall table of causes of hospital admissions in the public health network in Barreiras, accounting for 18% of all hospitalisations in the municipality, with 23,325 admissions over 11 years,

If we disregard the 31 per cent of hospitalisations due to pregnancy, childbirth and the puerperium[3] , because these are not actually illnesses, since they are related to pregnancy and the generation of life, morbidities due to diseases related to the respiratory system take first place in the

[3] Puerperium - the period of six to eight weeks after childbirth in which the woman experiences physical and psychological changes, tending to return to the state that characterised her before pregnancy.

total number of records.

Table 6 also shows that, over the period analysed, there was a decrease in the number of annual hospitalisations for ARDs. As a result, in 2010, Chapter X came third in the total number of hospitalisations, which may be linked to possible changes in the population's age distribution.

4.2.3. Incidence of respiratory diseases by age group

The incidence of respiratory diseases is not evenly distributed throughout the population, as they are more prevalent in certain age groups, which are considered to be more vulnerable, as stated by Neto (2005). Thus, age characterisation of the population is fundamental for identifying risk groups and for understanding the proportion and intensity of occurrence of respiratory diseases.

It's important to note that, as shown above in Graphs 12A and 12B, even though the total population grew by 5,578 inhabitants over the last 11 years, the population of children aged between 0 and 9, considered to be a group most vulnerable to ARDs, fell by 7,000 inhabitants.It was also observed that the group with the largest concentrated population, between the ages of 15 and 25, does not imply the group with the highest number of hospitalisations, since, according to Graph 13, the most affected population was the group of children and adolescents, which confirms that this is the group most predisposed to diseases of the respiratory system.

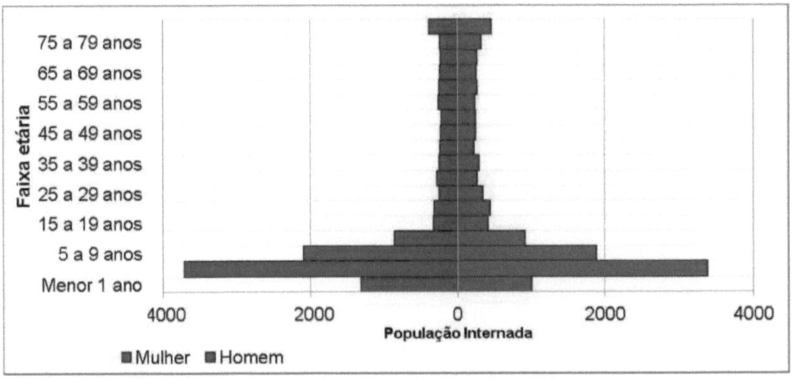

Graph 13: Morbidity by age group of residents in Barreiras-BA.
Source: DATASUS (2012)

This is also confirmed by Prietsch (2003), who investigated respiratory diseases in children and the influence of the domestic environment on their occurrence. It was also found that, even outdoors, this group is more sensitive to the behaviour of weather elements.

Graph 13 also shows the significant proportion of hospitalisations among children, where the group of people under the age of 1, with 2,332 hospitalisations, and people aged 1 to 10, with 11,082 hospitalisations, are the most affected by respiratory diseases, accounting for 57% of all hospitalisations, with 13,414 recorded in the 11 years of data consulted, according to DATASUS (2012).

The 1 to 4 age group had the highest proportion of hospitalisations per inhabitant, with a population of 9,423 in 2010, and 7,096 hospitalisations for this age group from 2000 to 2010, representing 75% of the population.

Therefore, with the confirmation of the high vulnerability of this group of children, there is a need to emphasise these age groups in greater detail, with more in-depth analyses and observations being carried out in the three initial age groups, i.e. the populations under 1 year old, 1 to 4 years old and 5 to 9 years old.

4.3 Association between atmospheric behaviour and respiratory diseases

The results of the analysis of hospital data on respiratory diseases and their association with meteorological behaviour were drawn up using the seasonal climate periods identified.Thus, the two seasonal periods identified in Barreiras confirm that the incidence of hospital morbidity due to respiratory diseases in the years analysed had a lower percentage in the rainy season, where 47% of the 17,707 hospital admissions recorded from January 2000 to December 2010 occurred in this period, while 53% of admissions occurred in the dry season, as can be seen in Graph 14.

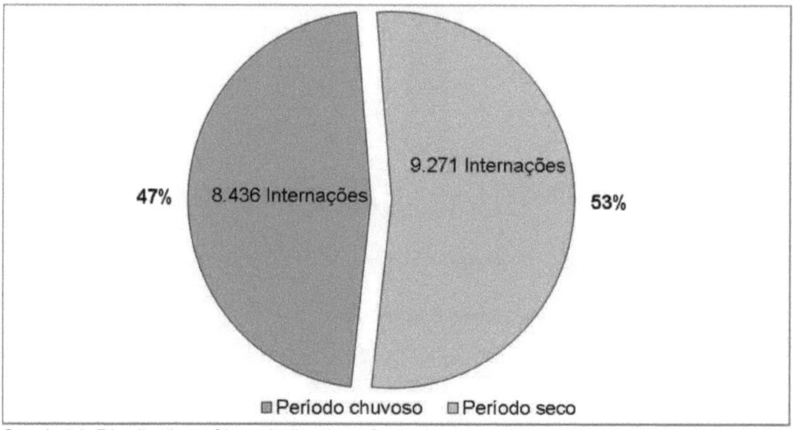

Graph 14. Distribution of hospitalisations for respiratory diseases based on the seasonality of the climate. Source: DATASUS (2012)

After seeing a higher number of hospitalisations in the dry season, as shown in Graph 14, it can be said that atmospheric characteristics attributed to the months of May to October, such as low

RH, significant thermal amplitudes and a visual increase in the concentration of particulate matter, could potentially be a determining factor in the increase in hospitalisations for respiratory diseases.The increase in hospitalisations is shown in Graph 15, which shows the central trend line of monthly hospitalisations for respiratory diseases in the municipality of Barreiras. The months with the highest relative growth compared to the previous month are March and April, with a 19% and 18% increase respectively, and November with a 29% increase in hospitalisations compared to October.

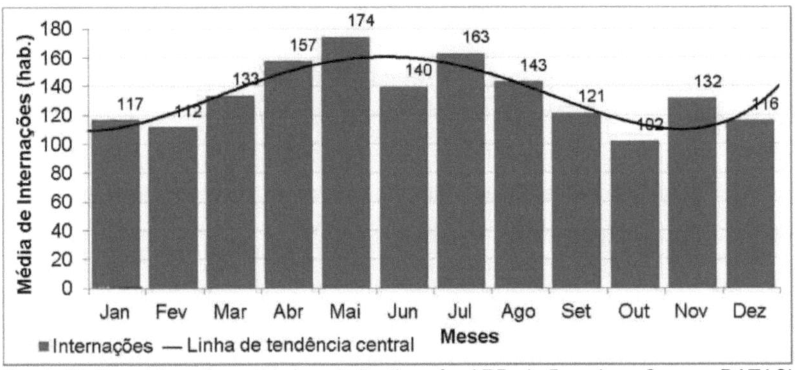

Graph 15: Trend line of monthly hospitalisations for ARDs in Barreiras. Source: DATASUS (2012)

This confirms the hypothesis raised in this study, since there are a greater number of hospitalisations in the period when the behaviour of the meteorological elements is a preponderant factor in the increase in the occurrence of ARDs.

The difference in the number of hospital morbidities between the dry and rainy periods, although not significant, with a 7% difference in the total number of hospitalisations due to ARDs from 2000 to 2010, makes it clear that the atmospheric characteristics in the dry and rainy months affect the number of hospitalisations.

This difference in the number of hospitalisations between the two seasonal periods can reach over 16%, as in 2005 and 2009, according to Table 7.

Table 7. Hospitalisations for respiratory diseases between dry and rainy periods

Percentages of Hospitalisations					
Year	Rainy season	Percentage	Dry Period	Percentage	Percentage Increase
2000	1.040,5	50%	1.040,5	50%	0%
2001	946	50%	950	50%	0%
2002	858	49%	905	51%	3%
2003	827	50%	827	50%	0%
2004	633	49%	657	51%	2%
2005	546	42%	752	58%	16%
2006	619	48%	681	52%	5%
2007	714	47%	801	53%	6%

2008	687	43%	905	57%	14%
2009	733	41%	1049	59%	18%
2010	692	45%	844	55%	10%
Total	1.406	47%	1.545	53%	7%

Source: DATASUS (2011)

However, Table 7 also shows that even though there is a higher incidence of hospitalisations in the dry season, seasonality alone cannot be considered a determining factor, since the number of hospitalisations in the hydrologically wet season is also relatively significant.

The monthly variation also confirms this finding, where the average monthly percentages shown in Graph 16 reveal that the transition months between the rainy and dry periods, from April to July, show a monthly percentage above 10% of all hospitalisations.

This means that there are months in both the dry and rainy periods with higher percentages of hospitalisations, especially in the transition from rainy to dry periods, as well as those with lower percentages of hospitalisations, such as October with 6%, while in the rainy periods there are months with a significant number of hospitalisations, such as April with 10%.

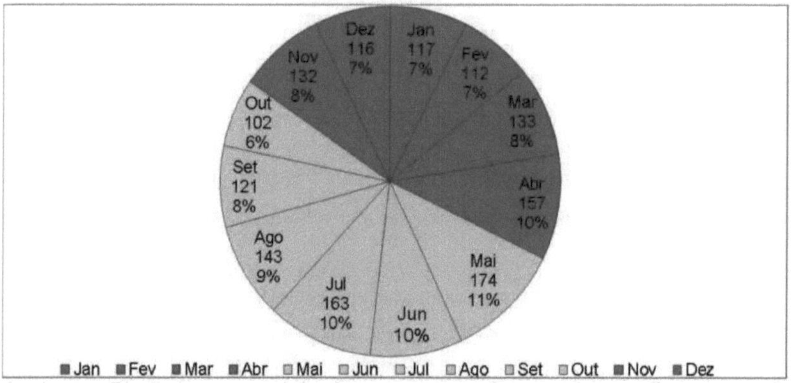

Graph 16. Monthly distribution of hospitalisations for respiratory diseases from 2000-2010 in Barreiras-BA. Source: DATASUS (2012).

The month of May saw the highest number of hospitalisations in general, with an average of 175 people suffering from respiratory diseases. This is the month of seasonal transition, as verified by the meteorological data, in which the greatest changes in atmospheric behaviour were found, such as a decrease in precipitation, RH and the consequent increase in the temperature range. It should also be noted that in the months of greatest water scarcity, there are many fires in the region, as shown in Figure 11 A, which highlights the hotspots in the dry period, confirmed in Figure 11B, which shows a significant decrease in the number of hotspots. This is mainly due to the expansion of pasture and plantation areas, as stated by Viganó (2011) in a study on the identification of hotspots using satellite images.

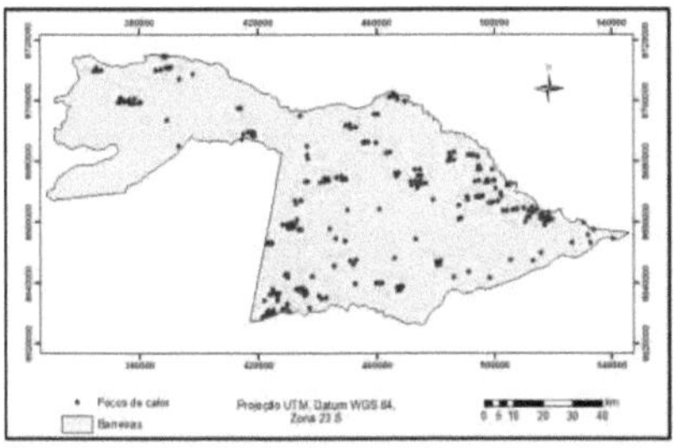

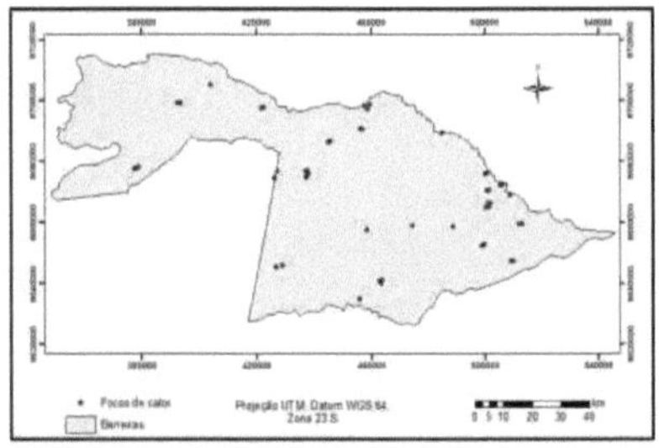

Figure 11A Heat spots during the dry season. Source: Viganó (2011).
Figure 11B Heat spots in the rainy season. Source: Viganó (2011).

The author also identified that many forest fires in the municipality occur mainly at the beginning of the dry season and are caused by anthropogenic activities linked to agriculture, such as clearing areas, renewing pastures, burning waste to eliminate pests and diseases, hunting techniques, and the "natural" fires of the Cerrado, caused by low humidity and high sunlight.

These fires release large quantities of gases and particulate matter into the atmosphere, leaving the air with a higher concentration of suspended particles, especially in the urban area of Barreiras, which makes it "heavy" and harmful to health.The intense incidence of fires close to the urban area proved to be a strong aggravating factor for the occurrence of respiratory diseases, when comparing the number of hospitalisations with the number of hotspots in the 5 km, 10 km, 15 km and 20 km buffers around the city's urban area, as shown in Table 8.

Table 8. Concentration of hotspots near the urban centre of Barreiras

	Heat sources					
Year	Urban sprawl	0 to 5km	5 to 10km	10 to 15 kilometres	15 to 20 kilometres	Total
2000	0	16	10	11	17	54
2001	1	29	23	19	40	112
2002	2	28	47	64	108	249
2003	3	56	34	67	130	290
2004	6	65	106	151	92	420
2005	3	76	104	174	168	525
2006	1	50	32	46	66	195
2007	10	98	231	364	314	1017
2008	77	108	39	77	108	409
2009	0	7	27	65	105	204
2010	10	132	170	227	370	909

Source: INPE (2012)

The years 2005, 2007, 2008 and 2010 had the highest numbers of hotspots, which can be associated with the percentage increase in hospitalisations shown in Table 7, which had the highest rates in the same period. Figure 12 shows the number of hotspots in these four years that had the highest number of fires close to the urban area.

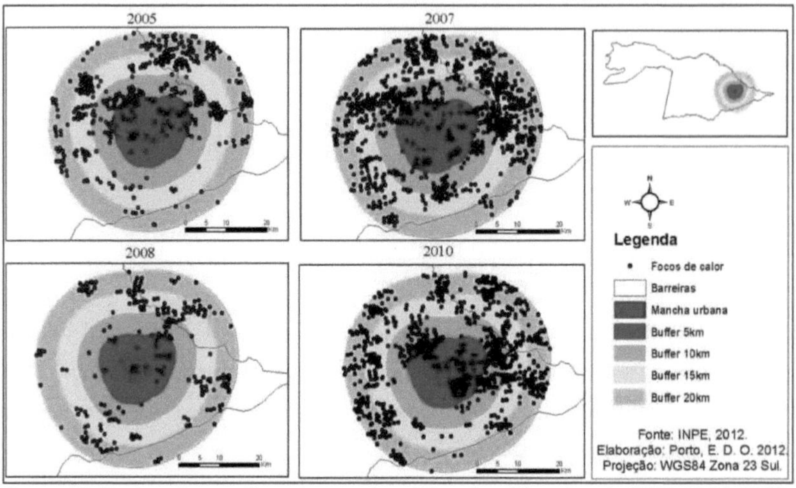

Figure 12. Heat spots near the urban centre of the city of Barreiras in 2005, 2007, 2008 and 2010. Source: INPE (2012).

Thus, forest fires are a major problem in the municipality, where Souza (2011) states that in Barreiras, in 2010, approximately 60,000 ha were burnt, an area equivalent to 7% of the municipality.

The photographic records of the urban atmosphere of Barreiras show that there is a daily oscillation in the concentration of particulate matter, which is also inversely proportional to temperature and relative humidity. This relationship is even more evident in the dry months, when the amount of PM present in the atmosphere is greater.

Particulate matter released into the city's atmosphere through aerosol smoke and CO2 leaves the air greyish and reduces visibility, which further reduces urban air quality, as shown in Figure 13.

Figura 13. Forest fire in Serra da Bandeira in the city of Barreiras on 29/08/2012. Photo: Porto, 2012.

Similarly, a practice that is very damaging to air quality and health, verified in the city through studies carried out by the Barreires Department of the Environment (2011), is urban fire, shown in Figures 14 and 15, a practice that has become commonplace, and used mainly to burn rubbish and household waste and to clear vacant lots, has become both a public health problem and a matter of environmental education.

Figura 14. Panoramic view of fires in the city of Barreiras on 13/05/2011. Photo: Porto, 2012

Figura 15. Burning to clear a vacant lot in the Morada Nobre neighbourhood in Barreiras on 20/09/2012. Photo: Porto, 2012

This problem becomes even more significant because, according to the studies mentioned above, the shape of the terrain that surrounds the city of Barreiras prevents the gases and particulate matter released into the atmosphere from being rapidly transported outside the urban environment.

In the hydrologically dry months, the particulate matter concentrated in the atmosphere leaves the air greyish in colour, as seen in Figure 17, which contrasts with the clear appearance seen in the atmospheric air during the rainy season in Figure 16.

Figure 16. Panoramic view of the urban atmosphere of Barreiras during the rainy season (11 December 2011). Photo: Porto (2011).

The smoke released into the air by fires and vehicles during dry periods, as shown in Figure 17, causes an increase in soot and particulate matter in the urban atmosphere, which, together with the low humidity and emissions of gases and aerosols, causes greater dryness of the nasal mucosa, according to the São Paulo Department of the Environment (2008), which favours the spread of respiratory diseases.

Figure 17. Panoramic view of the urban atmosphere of Barreiras during the dry season (11 October 2011). Photo: Porto (2011).

Factors related to a lack of infrastructure, such as unpaved streets and a lack of urban cleanliness, can also be associated with the decline in air quality. These are indicated by the National Health Surveillance Agency (ANVISA, 2009) as one of the main factors in the occurrence of respiratory diseases, since small particles of sand and silt deposited in the streets can be suspended by the wind and vehicles and inhaled.

These processes lead to an increase in soot and particulate matter in the urban atmosphere, which, together with low humidity and emissions of gases and aerosols from vehicles, cause the nasal mucosa to dry out more, according to SEMASP (2008), which favours the spread of respiratory diseases.

The increase in hospitalisations during periods of scarcity can also be intensified by anthropogenic factors such as forest and urban fires, the emission of aerosol gases and CO2, mainly from vehicles, which increase the particulate matter in the air and when inhaled can lower the body's immunity and make it more susceptible to illnesses such as colds and flu, which can develop into more serious chronic illnesses.

The comparison between the monthly data on hospital morbidities from respiratory diseases and the seasonal climate in the municipality of Barreiras showed that in the hydrologically dry period there was a significant increase in hospitalisations over the 11 years analysed (2000 to 2010), as shown by the trend line in Graphs 17A and 17B.

The number of hospitalisations increased by an average of 54% from the rainy season to the dry season, with the average number of annual hospitalisations in the wet season rising from 1,450 to 2,224 in the dry season.

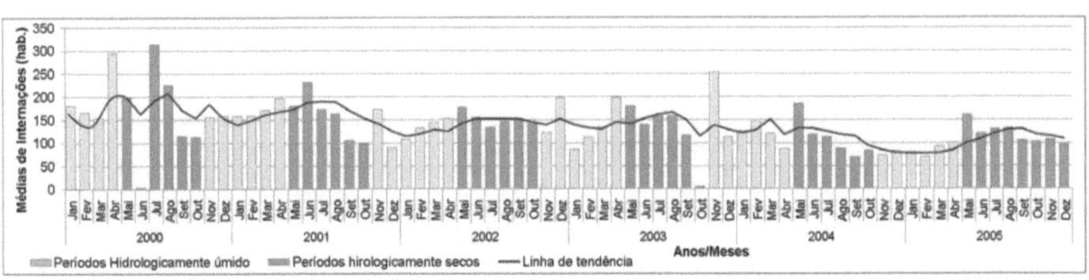

Graph 17A. Monthly hospital morbidity due to diseases of the respiratory system from 2000 to 2005. Source: DATASUS (2012)

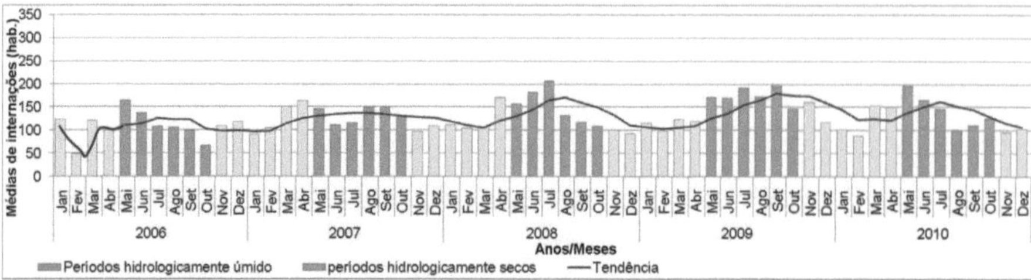

Graph 17B. . Monthly hospital morbidity due to diseases of the respiratory system from 2006 to 2010. Source: DATASUS (2012).

The data shown in Graphs 17A and 17B take into account the entire population and do not distinguish categories of greater vulnerability, such as age groups, which increases the risk of false positives for this association, since not all individuals who are affected by ARDs have an aggravation directly related to the weather, but to other conditions, such as smoking-related diseases or lifestyle, which are generally related to adults.

It should also be noted that there is a considerable difference in the degree of vulnerability and predisposition to illness for each age group, which forms the risk groups by age. Therefore, even though the data in Graphs 17A and 17B show that there is an association between respiratory illnesses and the seasonal climate, the children's age group is more sensitive to weather conditions, as shown in Graph 18, and can therefore be more easily affected by respiratory illnesses.

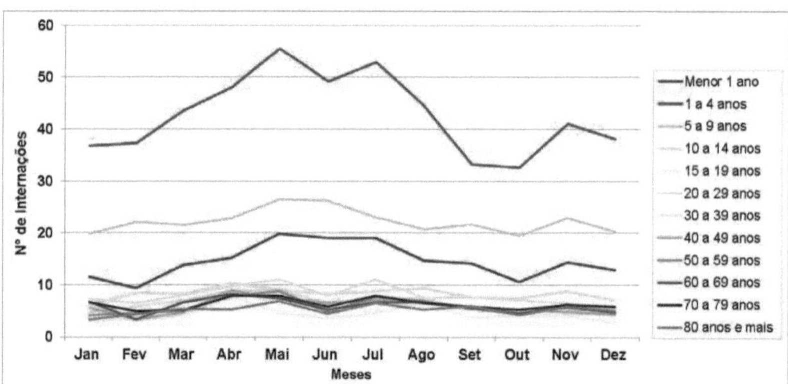

Graph 18. Average monthly hospitalisations by age group from 2000 to 2010. Source: DATASUS, 2012.

Having identified the greater vulnerability of the children's group (age groups under 1 year, 1 to 4 years and 5 to 9 years), it is necessary to emphasise the age groups of this group, since, according to Bakonyi (2004), the bodies of these individuals have not been able to develop the necessary immunological protection to avoid the illnesses that affect the respiratory tract.

Graph 18 shows that these age groups, of people under 9 years old, have the highest average monthly hospitalisations, which account for more than 58% of the total value of hospital morbidity due to ARDs.

Analysing the data for this population, as shown in Charts 19A and 19B, showed that there is a greater association between hospitalisations and seasonal periods for this group than in the similar analysis carried out for the whole population, shown in Charts 17A and 17B.

This symmetry between meteorological data and hospitalisation data confirms that children are more vulnerable to contracting diseases that affect the respiratory system as a result of unhealthy weather conditions than adults, even considering that underreporting[4] hospitalisation for adults is higher than for children.

Thus, analysing the group of children showed a positive association between the transitions between the seasonal periods, which could mean that the highest incidence of respiratory diseases in children occurs in the dry seasonal period, which could possibly also be related to the period with the highest concentration of particulate matter in the urban atmosphere and low relative humidity, as well as the highest monthly temperature variations.

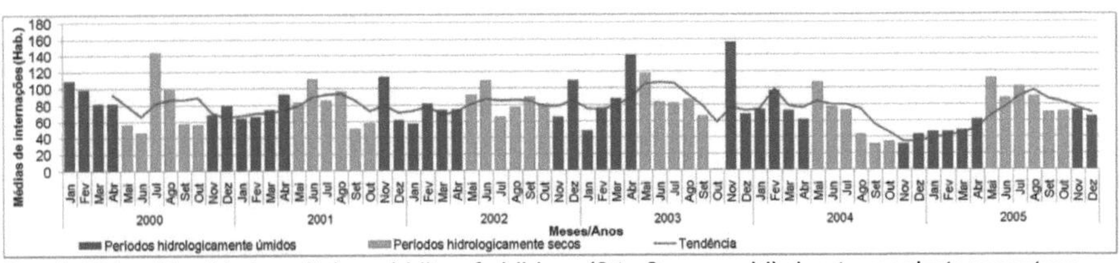

Graph 19A. . Monthly hospital morbidity of children (0 to 9 years old) due to respiratory system diseases from 2000 to 2005.
Source: DATASUS (2012).

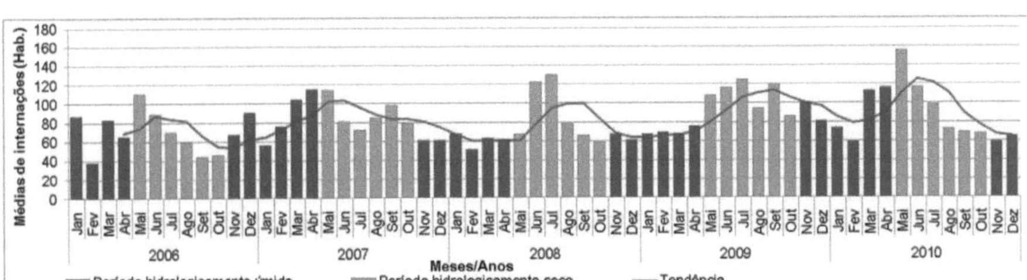

Graph 19A. . Monthly hospital morbidity of children (0 to 9 years old) due to respiratory system diseases from 2006 to 2010.
Source: DATASUS (2012).

Therefore, in the universe of data on children up to 9 years old, the 1 to 4 age group stands

[4] In epidemiology, underreporting is the difference between the number of illnesses caused by a given pathology and the number of notifications made about it (BRASIL, 2005). It is possible to say that underreporting is an indicator that access to health is a right that has been denied to a significant portion of the population (Conceição, 2003).

out as having the highest number of hospitalisations, with 31% of the total, followed by the 5 to 9 age group, with 16% of the total, and the age group of newborns, under 1 year old, with 11% of hospitalisations.

However, as shown in Table 9, the age group under 1 year old, with 2,337 individuals, stands out the most, because when compared to the age groups 1 to 4 and 5 to 9 years old, it has proportionally the highest percentage of hospitalisations by total number of individuals, with 7.49%.

Table 9. Proportion of hospitalisations among age groups.

Age group	Population	Average number of hospitalisations (per year)	Percentage
Under 1 year old	2.337	175	7%
1 to 4 years	9.423	513	5%
5 to 9 years	12.017	267	2%

Source: DATASUS, 2012.

Thus, hospital morbidity data for ARDs show a high prevalence[5] in children up to 1 year of age and may be related to a complex combination of factors, including breastfeeding. In this sense, Ostroswky (2001) suggests that breastfeeding can be a protective factor against more serious respiratory diseases. Therefore, the prevalence of respiratory diseases is concentrated in the autumn and winter months, especially from May to August, particularly in the under-5 age group (under 1 year old and 1 to 4 years old), who, proportionally, are the most vulnerable, as shown in Graphs 20A, 20B and 20C.

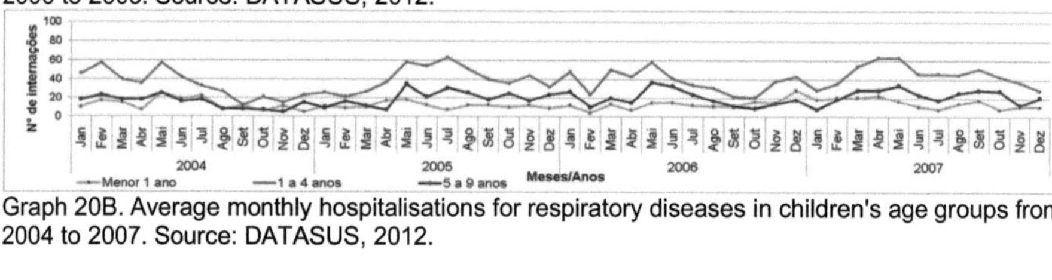

Graph 20A. Average monthly hospitalisations for respiratory diseases in children's age groups from 2000 to 2003. Source: DATASUS, 2012.

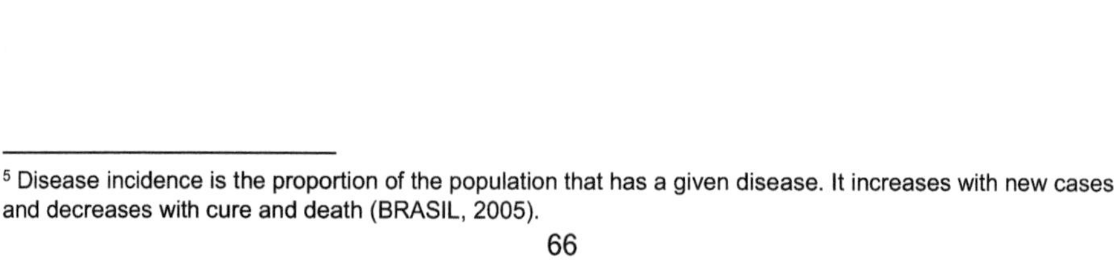

Graph 20B. Average monthly hospitalisations for respiratory diseases in children's age groups from 2004 to 2007. Source: DATASUS, 2012.

[5] Disease incidence is the proportion of the population that has a given disease. It increases with new cases and decreases with cure and death (BRASIL, 2005).

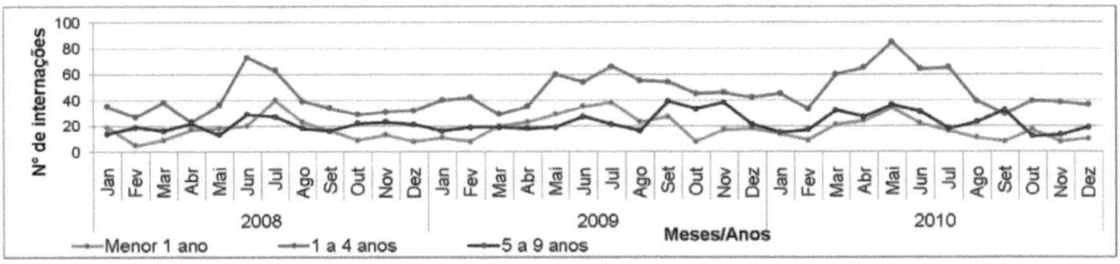

Graph 20C. Average monthly hospitalisations for respiratory diseases in children's age groups from 2008 to 2010. Source: DATASUS, 2012.

In addition to climatic seasonality, other important risk factors can explain the prevalence of respiratory diseases in children, since even in the hydrologically humid months, despite the proportional decrease, hospital morbidity remains high.

According to Brasil (2012), among the factors associated with a higher frequency and severity of episodes of acute respiratory infections in these age groups are malnutrition, a high incidence of low birth weight, lack of or short duration of breastfeeding, lack of immunisation through vaccines, contamination of household air due to the use of biomass fuels for heating and cooking food.

Low family income can also be mentioned, which is often associated with higher household density, the use of clothing unsuitable for the climate, factors related to a higher incidence of acute respiratory infections, and a lower level of maternal education.

The Ministry of Health also states that some factors and peculiarities linked to children with upper airway infections are recurrent from the environment, bad habits and inadequate environmental education, such as smoking at home, poor ventilation, humidity, mould, excessive contact with common allergens such as dust and mites.

CHAPTER 5

FINAL CONSIDERATIONS

In relation to the general objective of this research, we initially looked at the existence of an association between climatic behaviour and the occurrence of respiratory tract diseases, considering the average monthly values of the meteorological elements of precipitation, temperature, relative humidity and the records of hospital morbidity due to respiratory tract diseases.

When analysing the meteorological data, it was observed that the climate of Barreiras is divided into two seasonal periods, one rainy, lasting up to six months, and one dry, with six months or more of water scarcity, which are determined by the action of the air masses that act over Brazil.

In the period of water scarcity, or hydrologically dry, it was found that there was a positive result in the association between respiratory diseases and climate, when comparing the two variables: hospitalisations with meteorological data

Due to the limitations of hospitalisation data, such as the lack of information on daily figures, underreporting and the generalisation of diseases into chapters (ICD-10), this association did not allow us to state that seasonal atmospheric conditions are the only or main preponderant factors for the increase in hospitalisations.

However, it was noted that there is a complex intersection of both anthropogenic and natural factors, which in combination compromise environmental quality.

Thus, it was concluded that the occurrence of respiratory diseases may be much more linked to the conditions produced by the seasonal climate related to the environment, such as the increase in particulate matter, smoke and aerosols, than to meteorological factors.

The topic of respiratory diseases and the seasonal climate was grouped into: determinants and risk factors in the different age groups and utilisation of health services among children with respiratory problems, as this age group is the most frequent in this review.

Knowledge about the determinants and risk factors, the use and organisation of services, and the prevalence and incidence of respiratory disease in children is fundamental for improving the quality of care.

In this sense, it is very important to pay special attention to the children's age group (under 1 year old, 1 to 4 years old and 5 to 9 years old), because as well as having the highest percentage of hospitalisations, it can favour early detection, prevention of risks and damage and promotion of the quality of life of children and families.

It is also necessary to implement measures that can mitigate the impact of the season on the body of individuals, especially in environments that concentrate a greater number of children, such as schools and crèches, with the use of resources and equipment that provide greater comfort, such as well-ventilated and ventilated spaces, wooded outdoor areas, air humidifiers, indoor air conditioners and suitable clothing.

Points related to urban infrastructure can also be highlighted, such as urban afforestation, the construction of communal spaces and green areas, as well as the paving of urban roads, more frequent street cleaning, in addition to decisive actions to improve air quality, such as the more rigorous fight against forest fires in the municipality and burn-offs within the urban area, especially during dry periods.

These improvements in infrastructure and, consequently, in the quality of the urban atmospheric environment, can provide a better quality of life for the entire population and reduce the prevalence of respiratory diseases in children as well as the incidence of these diseases in adults, which could mean a long-term investment.

It is recommended that an air quality management department be set up within the Barreiras SEMA (Environment Department) to regulate industrial processes, motor vehicles, fires and other anthropogenic activities responsible for large emissions of PM and CO2 into the atmosphere, through regulations and municipal laws that can improve air quality, especially in urban areas.

Thus, this sub-secretariat must monitor air quality data using automatic and manual stations, analyse the data and disseminate it through daily bulletins and annual reports to the population, managers and decision-makers, universities, research and public health institutions and companies, enabling the formulation of public policies to improve air quality in the city and its surroundings.

CHAPTER 6

REFERENCES

AYOADE, J. O. **Introduction to climatology for the tropics.** 12ª Edition. Rio de Janeiro: Bertrand Brasil, 2007.

ABC of Health. Available at http://www.abcdasaude.com.br/ accessed on 08/11/2011.

BAKONYI, S. M. C. **Distribution of childhood respiratory diseases by gender in the municipality of almirante Tamandaré/PR.** Anais 12°Egal 2009.

BARCELLOS, C. **Geoprocessing, environment and health:** a possible union? Cadernos de Saúde Pública. Rio de Janeiro, v. 12, n. 3, p.389-397, jul./set. 1996.

BARCELLOS, C. **Georeferencing health data at the sub-municipal scale:** some experiences in Brazil. Epidemiologia e Serviço de Saúde [online]. Rio de Janeiro, 2008, vol.17, n.1, pp. 59-70.

BONFIM, C. **Epidemiology and Geography:** from the beginnings to geoprocessing. Revista Espaço para a Saúde, Londrina, 2008. v. 10. p. 53- 62.

BRAGA, A. L. F. **Association between air pollution and respiratory and cardiovascular diseases in the city of Itabira, Minas Gerais, Brazil.** Cad. Saúde Pública [online]. 2007, vol.23, p.p. 570-578.

BRAZIL, Ministry of Health. **Spatial approaches in public health.** Series B. Basic Health Texts. Brasília, 2006.

BRAZIL, Ministry of Health. **Epidemiological surveillance guide -** Series A. Norms and Technical Manuals. Health Surveillance Secretariat. DF, MS, 2005

BRAZIL. Centre for Weather Forecasting and Climate Studies. **BDBurns.** Available at http://www.dpi.inpe.br/proarco/bdqueimadas/. Accessed on 12/09/2012.

BRASIL. **City Statute: guide for implementation by municipalities and citizens.** 2 ed. Chamber of Deputies: Coordination of Publications. Brasília, 2002.

BRAZIL. Brazilian Institute of Geography and Statistics. **©Cities.** 2010. Available at: http://www.ibge.gov.br/cidadesat/topwindow.htm71. Accessed on 10/02/2012.

BRAZIL. Brazilian Institute of Geography and Statistics. 2011. Available at: http://www.ibge.gov.br/home/estatistica/populacao/censo2010/. Accessed on 03/05/2012.

BRAZIL. National Meteorological Institute. **INMET.** Available at: http://www.inmet.gov.br/portal/index.php?r=home/index. 11/09/2011.

BRAZIL. Ministry of Health / SUS. **Decentralisation of Health Actions and Services:** the audacity to comply with and enforce the law, 1993.

BRAZIL. Ministry of Health. **DATASUS.** Available at: **www.datasus.gov.br/.** Accessed on 11/09/2011.

BRAZIL. Ministry of Cities. **GEOSNIC.** 2012. Available at: www.cidades.gov.br/ accessed *on 05/05/2012.*

BRAZIL. Ministry of Mines and Energy. **CONPET: National Programme for the Rationalisation of the Use of Petroleum Derivatives and Natural Gas.** Petrobras - Rio de Janeiro, 2011. Available at http://www.conpet.gov.br/download/pdf/economizar_diesel_saude.pdf Accessed on 20/02/2012.

GAMARA, G. **Spatial analysis and geoprocessing.** In: Druck, S.; Carvalho, M.S.; Câmara, G.; Monteiro, A.V.M. (eds). **Spatial Analysis of Geographic Data.** EMBRAPA, Brasília, 2004.

CÂMARA, G. and DAVIS, C. **Introduction to geoinformation science.** INPE. São José dos Campos, 2001.

CÂMARA, G. *Geocomputation techniques for spatial analysis: are they relevant to health data?* Caderno de Saúde Pública, Rio de Janeiro, 2001.

CARNEIRO, E. O. **The use of geoprocessing techniques in public health:** spatial analysis applied to determine areas of endemic diseases. Proceedings of the x SBSR. Foz do Iguaçu, 2003.

CONCEIÇÃO, C. L. **Relationships between climatic elements and respiratory diseases in Rio Grande do Sul.** Master's dissertation. Porto Alegre, 2003.

CONFALONIERI, U. E. C. **Climate variability, social vulnerability and health in Brazil.** Terra Livre Magazine: Global and Local Climate Change Year 19 - V. 1- Oswaldo Cruz Foundation. Rio de Janeiro, 2002. p.p. 192-204

CORRÊA, R. L. **O Espaço Urbano.** Editora Ática, Série Princípios, 3° edition, n°174, 1995. p.1-16.

COSTA, G. F. **Geoprocessing:** Use and Application in Public Health and Environmental Health. In. 1st Meeting of the National Association for Postgraduate Studies and Research in Environment and Society. ANPPAS. Indaiatuba, 2002.

COSTA, H. S. de M. **Desenvolvimento urbano sustentável:** Uma contradição de termos? 8° Encontro Nacional da Anpur. Porto Alegre, mimeo, 1999.

COSTA, M. C. N.; TEIXEIRA, M.G.L.C. **The concept of "spaces" in epidemiological research.** Cadernos de Saúde Pública. Volume 15, No. 2, April/June, Rio de Janeiro, 1999.

COVISA, Health Surveillance Coordination. **Technical Report on Low Air Humidity.** São Paulo Municipal Health Department. São Paulo, 2010.

CZERESNIA, D. **The concept of Space in Epidemiology:** A historical and epistemological interpretation. In: Cadernos de Saúde Pública. Rio de Janeiro, 2000.VoL 16, p.595- 605.

DEFFUNE, G. **Influence of cloudiness on insolation values.** Maringá, 1993.

Health spending. São Paulo - Uniesp p.27-46, 2006.

GOLDEMBERG, J. **Energia, meio ambiente e desenvolvimento.** EDUSP - São Paulo, 2003.

GURGEL, H. C. **New sensors and their benefits for studies related to human health.** Proceedings of the 13th Brazilian Symposium on Remote Sensing. Florianópolis- INPE, 2007. p. 3961-3966.

HYGEIA. Brazilian Journal of Medical and Health Geography. Available at: HTTP://www.hygeia.ig.ufu.br/. Accessed on 13/02/2012.

JESUS, E. F. R. **Interface between Climatology and Epidemiology:** a geographical approach. GeoTextos. Salvador, 2010. vol. 6, p.p. 211-236.

JUNQUEIRA, R. D. **Medical Geography and Health Geography.** HYGEIA, [online], 2009.

MARTINS, R. A. **Contágio: história da prevenção das doenças transmissíveis.** São Paulo: Moderna, 1997. Available on HTTP://www.ifi.unicamp.br/~ghtc/Contagio/ on 5/02/2012.

MORAIS, R. H. S. G. **A geografia médica e as expedições francesas para o Brasil:** uma descrição da estação naval do Brasil e da Prata (1868-1870). Manguinhos, Rio de Janeiro, 2007. p.p.39-62.

OSTROSWKY, R. **Respiratory diseases in children from 0 to 12 months and breastfeeding.** São Paulo. Unicamp -2001.

PEIXOTO, A. **Clima e saúde.** São Paulo: Cia. Editora Nacional, 1938. Available at HTTP://www.brasiliana.com.br/obras/clima-e-saude-introducao- biogeografica-a-civilizacao-brasileira. Accessed: 5/02/2012.

PINA, M. F. **Basic concepts of Geographic Information Systems and Cartography applied to health.** PAHO. Brasilia, 2000.

RIBEIRO, E. A. W. **Efficiency, effectiveness and efficacy in the planning of the**

ROJAS, L. I. *Geography* **and** *Health. Between histories, realities and utopias.* National Symposium on Health Geography. Presidente Prudente, 2003.

ROJAS, L. I. *Geography and health: themes and perspectives in Latin America.* In: Caderno de Saúde Pública. Rio de Janeiro, vol. 14, n°. 4, p.701-711, 1998.

SANTANA, P. **Saúde Urbana / Saúde dos Urbanistas. The lost paradigm of urbanity?** Portuguese Association for the Promotion of Public Health (APPSP). Lisbon, 2006.

SANTOS, M. **Brazilian urbanisation.** São Paulo: Hucitec, 1994.

SANTOS, **M. Por uma Geogradia nova.** São Paulo: University of São Paulo Press, 2008.

SIBBONS, H. L. **A contribuition to the study of potential evapotrenspiration.** Geografisker annaler. City 1962.

SOUZA, C. G. and S AN T' AN NA NETO, J. L. **Spatial distribution of respiratory diseases in medium-sized cities in Western São Paulo - Brazil:** a multi-causal analysis as a contribution to improving the quality of urban life. Paulista State University. São Paulo, 2007.

SOUZA, C. G. e SANT'ANNA NETO, J. L. **Geografia da saúde e climatologia médica: ensaios sobre a realção clima e vulnerabilidade,** HYGEIA[online] 2008.

SOUZA, C. G. e S AN T' AN NA NETO, J. L. **Geografia da Saúde e Climatologia Médica:** ensaios sobre a relação clima e vulnerabilidade.

SOUZA J-A. **Applied hydrology.** 2005. Available at http://pt.scribd.com/doc/50282292/Cap-1 - Ciclo-Hidrologico-2005. Accessed on 08/03/2012.

STEINBERGER, M. **The (Re)Construction of Myths about the (In)Sustainability of Urban Space.** Brazilian Journal of Urban and Regional Studies. A.3, n.4. p. 9 - 32. UFPE - Pernambuco, 2001.

TORRES, F. T. P. **Introduction to climatology.** Cengage Learning. São Paulo, 2011.

VENTURI, L. A. B. **Praticando a Geografia:** técnicas de campo e laboratório em geografia e analise ambiental. São Paulo, Oficina de Textos, 2005.

WHOQOL - **The World Health Organisation Quality of Life.** Available at: http://www.who.int/substance_abuse/research_tools/whoqolbref/en. Accessed on 03/08/2010:

YUNES, J. **Geographic Information Systems and health management in the municipality.** International Health Information Network. Rio de Janeiro, 1997.

CHAPTER 7

ANNEXES

Burning of a vacant lot in the Sandra Regina neighbourhood seen from the Serra do Mimo on 18/06/2011. Photo: Porto, 2011.

Sandstorm in the city of Barreiras seen from the Serra do Mimo on 28/01/12. Photo: Porto, 2012.

Fire in the Serra do Mimo on 30/08/2012 near houses in the Morada da Lua neighbourhood. Photo: Porto, 2012.

Fire in the Serra do Mimo on 29/08/2012 near the Jardim Imperial neighbourhood. Photo: Porto, 2012.

Panoramic view of the city of Barreiras on 13 September 2011 at 7:08 am. Photo: Porto, 2011.

Panoramic view of the city of Barreiras on 13 September 2011 at 4:50 pm. Photo: Porto, 2011.

View of the city of Barreiras at the junction of the BR242 motorway and the Barreiras Airport runway. Photo, Porto, 2011.

Significant layer of PM in the urban area of Barreiras, seen from the intersection of BR242 and the airport runway on 12/07/2011. Photo, Porto, 2011.

MIX
Papier aus verantwortungsvollen Quellen
Paper from responsible sources
FSC® C105338

Printed by Books on Demand GmbH, Norderstedt / Germany